Electrochemical Engineering

Principles

PRENTICE HALL INTERNATIONAL SERIES
IN THE PHYSICAL AND CHEMICAL ENGINEERING SCIENCES

NEAL R. AMUNDSON, SERIES EDITOR, *University of Houston*

ADVISORY EDITORS

ANDREAS ACRIVOS, *Stanford University*
JOHN DAHLER, *University of Minnesota*
THOMAS J. HANRATTY, *University of Illinois*
JOHN M. PRAUSNITZ, *University of California*
L. E. SCRIVEN, *University of Minnesota*

BALZHIZER, SAMUELS, AND ELLIASSEN *Chemical Engineering Thermodynamics*
BUTT *Reaction Kinetics and Reactor Design*
CROWL AND LOUVAR *Chemical Process Safety*
DENN *Process Fluid Mechanics*
FOGLER *Elements of Chemical Reaction Engineering, 2nd edition*
HIMMELBLAU *Basic Principles and Calculations in Chemical Engineering,*
    5th edition
HINES AND MADDOX *Mass Transfer*
HOLLAND *Fundamentals and Modeling of Separation Processes*
HOLLAND AND ANTHONY *Fundamentals of Chemical Reaction Engineering,*
    2nd edition
KYLE *Chemical and Process Thermodynamics*
LEVICH *Physiochemical Hydrodynamics*
LUYBEN AND WENZEL *Chemical Process Analysis*
MODELL AND REID *Thermodynamics and Its Applications, 2nd edition*
NEWMAN *Electrochemical Systems, 2nd edition*
OLSON AND SHELSTAD *Introduction to Fluid Flow and the Transfer of Heat and Mass*
PRAUSNITZ, LICHTENTHALER, AND DE AZEVEDO *Molecular Thermodynamics of*
    Fluid-Phase Equilibria, 2nd edition
PRAUSNITZ ET AL. *Computer Calculations for Multicomponent Vapor-Liquid and*
    Liquid-Liquid Equilibria
PRENTICE *Electrochemical Engineering Principles*
RAMKRISHNA AND AMUNDSON *Linear Operator Methods in Chemical Engineering*
RHEE ET AL. *First-Order Partial Differential Equations: Volume I, Theory and*
    Application of Single Equations
RHEE ET AL. *First-Order Partial Differential Equations: Volume II, Theory*
    and Application of Hyperbolic Systems of Quasilinear Equations
SCHULTZ AND FAKIROV, eds. *Solid State Behavior of Linear Polyesters and*
    Polyamides
STEPHANOPOULOS *Chemical Process Control*
WHITE *Heterogeneous Catalysis*

Electrochemical Engineering Principles

Geoffrey Prentice

The Johns Hopkins University

Prentice Hall, Englewood Cliffs, New Jersey 07632

Library of Congress Cataloging-in-Publication Data

Prentice, Geoffrey.
   Electrochemical engineering principles / Geoffrey Prentice.
      p.   cm. -- (Prentice-Hall international series in the physical
   and chemical engineering sciences)
   Includes bibliographical references.
   ISBN 0-13-249038-2
   1. Electrochemistry, Industrial.   I. Title.   II. Series.
TP255.P74   1991
660'.297--dc20   90-35448
                                                            CIP

Editorial/production supervision: Barbara Marttine
Manufacturing buyer: Kelly Behr

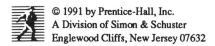 © 1991 by Prentice-Hall, Inc.
A Division of Simon & Schuster
Englewood Cliffs, New Jersey 07632

Printed in the United States of America
10 9 8 7 6 5 4 3 2 1

ISBN   0-13-249038-2

Prentice-Hall International (UK) Limited, *London*
Prentice-Hall of Australia Pty. Limited, *Sydney*
Prentice-Hall Canada Inc., *Toronto*
Prentice-Hall Hispanoamericana, S.A., *Mexico*
Prentice-Hall of India Private Limited, *New Delhi*
Prentice-Hall of Japan, Inc., *Tokyo*
Simon & Schuster Asia Pte. Ltd., *Singapore*
Editora Prentice-Hall do Brasil, Ltda., *Rio de Janeiro*

To Merlyn, Loren, Aline

"My desire [was] to escape from trade, which I thought vicious and selfish, and to enter into the service of Science, which I imagined made its pursuers amiable and liberal..."

Michael Faraday, 1829

Contents

Preface

Electrochemical processes and devices have become widespread over the last century. Development of electrochemical technology paralleled understanding of basic phenomena during a period of rapid development in the 19th century. Batteries, fuel cells, plating, and the electrolytic production of aluminum and chlorine all evolved in the late 1800s. Today, products based on electrochemical phenomena account for tens of billions of dollars in the US.

Although electrochemical technology constitutes an important segment of the economy, a larger segment is dominated by the presence of abundant and relatively low-cost hydrocarbon sources. Over three-quarters of the energy in the US is derived from hydrocarbons, and hydrocarbon-based polymeric materials are ubiquitous in modern society. Because of these economic and industrial realities, academic curricula treating chemical process industries emphasize organic over inorganic chemical processes, and standard textbooks in thermodynamics, phase equilibria, kinetics, and mass transport reflect an organic chemistry emphasis.

When I developed the first course in electrochemical engineering here at Hopkins in the early 1980s, I relied on notes that I developed and several texts. Because the texts were directed toward different audiences, I found difficulty in reconciling the differences in conventions, nomenclature, and level of material. This book is an outgrowth of the notes developed for the electrochemical engineering course. In this book I emphasize the principles of electrochemical engineering at a level suitable for seniors and graduate students in chemical engineering.

In the past, one learned a modicum of electrochemistry in physical chemistry courses, which usually spanned the entire junior year. In modern chemical engineering curricula, physical chemistry is often reduced to one or fewer semesters, and electrochemistry is treated in one or two weeks. Because of this reduced emphasis on electrochemical processes, I review the basic principles of electrochemistry in Chapter 2. Simultaneously, I introduce a series of conventions and calculation procedures, that are used in the remaining chapters.

Thermodynamics, phase equilibria, kinetics, and mass transport are all topics familiar to the undergraduate chemical engineer. Because of the emphasis on organic chemistry noted above, practicing chemical engineers commonly deal with processes such as distillation, catalytic cracking, and reactions of hydrocarbon feedstocks. The most popular textbooks addressing chemical engineering fundamentals mirror the dominance of hydrocarbon processes and avoid the topics of cell thermodynamics, ionic phase equilibria, electrode kinetics or ionic mass transport. I include this subject matter in Chapters 3 to 6.

Chapter 7 represents a synthesis of fundamental concepts. Techniques for modeling an electrochemical system are presented along with mathematical methods for determining potential and current distributions for several important classes of problems. Emphasis is on a preliminary analysis of a cell to separate essential phenomena from second order effects.

A number of texts have been written on experimental techniques in electrochemistry. Rather than survey the literature, I discuss several common methods of use to the engineer. The rotating disk electrode is a popular tool in the electrochemical laboratory, and I discuss its utility in some detail. Because the characterization of mass transfer rates is often important, other experimental techniques having well-characterized hydrodynamics and mass transfer rates are also discussed.

The final chapter is an eclectic collection of current electrochemical systems and those under development. In the examples, I attempt to illustrate the application of principles to system design. There have been several notable advances that have dramatically improved the performance of existing processes, e.g., the chlor-alkali process. In other cases, e.g., fuel cells and electric vehicles, widespread adoption awaits advances in materials, design, or catalysis. I have chosen a few examples of systems under development to highlight areas needing additional research effort.

Most of the material in this text represents a logical extension and elaboration of areas in chemical engineering curricula. Readers with backgrounds in materials science, biomedical engineering, environmental engineering, and chemistry should also be familiar with the basic material. The mathematical treatment of ionic mass transport and current distribution requires a knowledge of ordinary differential equations. Although partial differential equations are introduced, a working knowledge of solution techniques is not required.

The material in this book can usually be covered in a one semester course. Several subjects can be eliminated with little loss of continuity. Some of these dispensable subjects include the sections on Debye-Hückel theory, mechanistic studies, and numerical methods.

Most realistic simulations of electrochemical cells require computer techniques. Several of the problems in Chapters 6 to 9 are best solved with the aid of a computer; however, the sections requiring a knowledge of computer skills can be omitted with no loss of continuity. I have included two Fortran programs, illustrating the computation of potential distributions in one- and two-dimensional cells.

A number of topics that I chose to include or exclude largely reflect personal taste. Probably, the most conspicuous inclusion is a treatment of the thermodynamics and kinetics of corrosion. I have several reasons for considering these topics. Corrosion is, in economic terms, the most important electrochemical process. It is a \$150 billion growth industry and should not be ignored by the practicing engineer. The principles pertaining to corrosion are rather general and apply to related fields. For example, the Pourbaix diagram is used by corrosion engineers as well as by environmental engineers and geochemists.

In addition, I chose to review several theories that are of more pedagogic than practical value. The Debye-Hückel theory and Stokes theory of ion conduction fall into this category. I included them because they provide a simple, yet quantitative, view of the essential phenomena. A number of topics that are arguably important were not included, among them ion transport in membranes, fluidized bed reactors, solid electrolytes, concentrated solution theory, molten salt electrolytes, semiconductor electrodes, and electrochemical impedance spectroscopy. I also chose not to include topics that are important in system design but are covered in a typical chemical engineering curriculum. Here, I have in mind areas such as heat transfer and process control.

I would like to thank colleagues and advisers who were influential in the development of the ideas in this book. At Berkeley, Charles

Tobias provided a rich and varied intellectual environment in which I conducted my doctoral work. His guidance and insight—scientific, economic, culinary, religious, and philosophical—have been invaluable. His comments and criticisms of portions of early drafts of the book were extremely useful in guiding revisions. The careful reading of the manuscript by Dan Gibbons, a teaching assistant in Charles Tobias's course, also contributed a fresh perspective. John Newman was influential as a teacher. He introduced us to the aesthetics and delights of rigor; he dazzled graduate student and colleague alike with his analytical brilliance. Rolf Muller, as co-chairman of our research seminars, introduced us to a fascinating array of experimental techniques. A number of the problems and examples in this book are based on courses, seminars, and research results from Berkeley. Former co-workers Paul Sides and Karrie Hanson helped, through extended discussions, in the clarification of many concepts in this book.

The editors at Prentice Hall were most helpful in guiding me through the various phases of production. The executive editor Michael Hays encouraged me during the initial stages of production. Barbara Marttine, the senior editor, did a fine job of massaging the prose and attending to the details through the final copy.

Here at Hopkins, Jerry Kruger and Pat Moran contributed to my appreciation of the electrochemical aspects of corrosion phenomena. Our cooperative ventures in teaching and research broadened my perspectives in this area. Many facets of fundamental electrochemistry were brightened by Eliezer Gileadi's numerous, but all too short, visits. Assistance from graduate students, postdoctoral fellows, and undergraduates, especially those who labored in courses with early drafts of the text, is gratefully acknowledged. I also thank those sponsors who have supported my work: NSF, NASA, ONR, Baltimore Gas and Electric, the Alcoa Foundation, and DuPont.

Geoffrey Prentice

General References

Electrochemical Engineering

T. Z. Fahidy, *Principles of Electrochemical Reactor Analysis* (New York: Elsevier, 1985).

E. Heitz and G. Kreysa, *Principles of Electrochemical Engineering* (New York: VCH, 1986).

F. Hine, *Electrode Processes and Electrochemical Engineering* (New York: Plenum Press, 1985).

Electrochemical Reactors: Their Science and Technology, M. Ismail, ed., (New York: Elsevier, 1989).

J. S. Newman, *Electrochemical Systems* (Englewood Cliffs, NJ: Prentice-Hall, 1973).

I. Roušar, K. Micka, and A. Kimla, *Electrochemical Engineering* (New York: Elsevier, 1986).

Tutorial Lectures in Electrochemical Engineering and Technology, vol. 79, no. 229, R. C. Alkire and D-T. Chin, eds., (New York: American Institute of Chemical Engineers, 1983).

Tutorial Lectures in Electrochemical Engineering and Technology, vol. 77, no. 204, R. C. Alkire and T. R. Beck, eds., (New York: American Institute of Chemical Engineers, 1981).

Electrochemistry

M. M. Baizer and H. Lund, *Organic Electrochemistry: An Introduction and a Guide* (New York: Marcel Dekker, 1983).

J. O'M. Bockris and A. K. N. Reddy, *Modern Electrochemistry* (New York: Plenum Press, 1973).

H. Rickert, *Electrochemistry of Solids* (New York: Springer Verlag, 1982).

P. H. Rieger, *Electrochemistry* (Englewood Cliffs, NJ: Prentice-Hall, 1987).

Series Publications

Comprehensive Treatise of Electrochemistry, J. O'M. Bockris et al., eds. (New York: Plenum Press, 1980-).

Modern Aspects of Electrochemistry, J. O'M. Bockris et al., eds. (New York: Academic Press, 1954-).

Advances in Electrochemistry and Electrochemical Engineering, C. W. Tobias, et al., eds. (New York: Wiley-Interscience, 1961-).

Techniques of Electrochemistry, E. Yeager and A. J. Salkind, eds. (New York: Wiley-Interscience, 1972-).

Ch. 4, Phase Equilibrium

A. L. Horvath, *Handbook of Aqueous Electrolyte Solutions* (Chichester, England: Ellis Horwood Ltd., 1985).

R. A. Robinson and R. H. Stokes, *Electrolyte Solutions* (London: Butterworths Scientific Publications, 1959).

Ch. 5, Electrode Kinetics

J. Albery, *Electrode Kinetics* (Oxford: Clarendon Press, 1975).

K. J. Vetter, *Electrochemical Kinetics* (New York: Academic Press, 1967).

Ch. 7, Modeling and Simulation

M. E. Davis, *Numerical Methods and Modeling for Chemical Engineers* (New York: John Wiley and Sons, 1984).

Ch. 8, Experimental Methods

W. J. Albery, and M. L. Hitchman, *Ring-disc Electrodes* (Oxford: Clarendon Press, 1971).

A. J. Bard and L. R. Faulkner, *Electrochemical Methods: Fundamentals and Applications* (New York: John Wiley and Sons, 1980).

E. Gileadi, E. Kirowa-Eisner, and J. Penciner, *Interfacial Electrochemistry: An Experimental Approach* (Reading, PA: Addison-Wesley Publishing Co., 1975).

Reference Electrodes, D. J. Ives and G. J. Janz, eds. (New York: Academic Press, 1961).

D. D. Macdonald, *Transient Techniques in Electrochemistry* (New York: Plenum Press, 1977)

D. T. Sawyer and J. L. Roberts, Jr., *Electrochemistry for Chemists* (New York: John Wiley and Sons, 1974).

Ch. 9, Applications

M. G. Fontana, *Corrosion Engineering* (New York: McGraw-Hill, 1986).

Industrial Electrochemical Processes, A. T. Kuhn, ed. (Amsterdam, NY: Elsevier, 1971).

Handbook of Batteries and Fuel Cells, D. Linden, ed. (New York: McGraw-Hill, 1984)

D. Pletcher, *Industrial Electrochemistry* (New York: Chapman and Hall, 1982).

Nomenclature

a	activity
A	Helmholtz free energy (J)
B	Tafel slope (mV/decade)
c	concentration (mol/cm^3)
C	capacitance (μF/cm^2)
d	diameter (cm)
D	diffusivity (cm^2/s)
D	dielectric constant
e	unit charge, 1.6×10^{-19} C/chg
E	potential (V)
$\mathbf{E_f}$	electric field (V/cm)
f	fugacity (atm)
F	Faraday's constant, 96,500 C/equiv
g	acceleration of gravity, 980 cm/s
G	Gibbs free energy (J)
Gr	Grashof number, $g(\rho_\infty - \rho_0)L^3/\rho_\infty \nu^2$
H	enthalpy (J)
i	current density (A/cm^2)
i_0	exchange current density (A/cm^2)
I	current (A)
I	ionic strength (mol/kg)
k	kinetic constant (cm/s)(cm^3/mol)$^{p-1}$
k_m	mass transfer coefficient (cm/s)
L	characteristic length (cm)
K	Equilibrium constant
m	molality (mol/kg)
m	mass (g)
M	molarity (mol/L)
M	molecular weight (g/mol)
M	symbol for chemical formula

n	number of electrons involved in a reaction
n	number of moles (only in partial derivatives)
N	flux (mol/s-cm^2)
N	number of moles
p	reaction order
P	pressure (atm)
q	charge density (C/cm^2)
q	heat transferred (J)
Q	charge (C)
r	reaction rate (mol/s-cm^2)
r	radius (cm)
R	universal gas constant, 8.31 J/mol-K
R	homogeneous reaction rate $(\text{mol/cm}^3\text{-s})$
R	resistance (ohm)
Re	Reynolds number, $dv\rho/\mu$
s	stoichiometric coefficient
S	entropy (J/K)
Sc	Schmidt number, ν/D
Sh	Sherwood number, $i_l L/nFDc_\infty$
t	transference number
t	time (s)
T	temperature (K)
u	ionic mobility $(\text{cm}^2\text{-mol/J-s})$
U	internal energy (J)
v	velocity (cm/s)
V	volume (cm^3)
V	cell potential (V)
w	work (J)
z	charge number

Greek Symbols

α transfer coefficient

α Debye-Hückel constant $(kg/mol)^{1/2}$

β symmetry factor

β coefficient of linear kinetic expression (mV)

γ molal activity coefficient

δ Nernst diffusion layer thickness (cm)

ϵ permittivity $(C^2/N\text{-}m^2)$

ϵ_0 permittivity of free space, $8.85 \times 10^{-12}\ C^2/N\text{-}m^2$

ϵ efficiency

η overpotential (V)

κ conductivity $(ohm^{-1}\text{-}cm^{-1})$

λ Debye length (Å)

λ ionic equivalent conductance $(cm^2/ohm\text{-}equiv)$

Λ equivalent conductance $(cm^2/ohm\text{-}equiv)$

μ electrochemical potential (J/mol)

μ viscosity (g/cm-s)

ν number of ions into which a species dissociates

ν kinematic viscosity (cm^2/s)

ρ density (g/cm^3)

ϕ electric potential (V)

ω rotation rate (rad/s)

Superscripts

$\ddagger$ activation

$*$ dimensionless

—   partial molar

$\cdot$ rate

0 standard state

$chem$ chemical

el electric

ec electrochemical

Subscripts

$\pm$	mean
$-$	anion
∞	bulk
$+$	cation
a	anodic
avg	average
c	cathodic
cn	concentration
$corr$	corrosion
f	formation
i	species type
j	species type
l	liquid junction
l	limiting
0	surface
O	oxidized species
r	reference
r	reduced
R	reduced species
rev	reversible
s	surface
w	working
x	oxidized

Chapter 1

Introduction

Electrochemical processes have been employed for over a century in industrial electrolysis, energy conversion, and metal deposition. An electrochemical route is usually chosen for one or more of its inherent advantages: energy efficiency, low temperature operation, ease of control, and low pollutant production.

1.1 Major Applications

Aluminum reduction and chlorine production are examples of processes that were initially carried out by ordinary chemical routes but were subsequently supplanted by more efficient electrochemical processes. In economic terms, these two processes are the most important, with a product market value of about $10 billion in the US; together they consume over 6% of the total domestic electrical output. Reduction of other metals from ores, electroörganic syntheses, and electroplating are also carried out on a large scale.

The economic values of major electrochemical processes and products, estimated in a National Research Council study [1], are summarized in Table 1.1. Developments in electrochemical processing and device fabrication are expected to lead to new markets in microelectronic devices, fuel cells, sensors, and corrosion control.

The beauty of electrochemical devices is that the energy of chemical bonding is converted directly to electrical energy. Because electrochemical energy conversion is not based on the transfer of heat

1

Process or Product	Annual Market ($ billion)
Aluminum	4
Sodium Hydroxide	3
Chlorine	2
Copper (electrolytic)	2
Other metals and chemicals	2
Electroplating	10
Batteries	4
Semiconductor processing	1
Total	28

Table 1.1: Major products or processes based on electrochemical technology.

between a hot and cold reservoir, Carnot limitations are avoided and inherently more efficient processes are, in principle, possible.

Many smaller niches are filled by electrochemical processes or devices. Machining a narrow cavity in a hard alloy may be impossible by ordinary mechanical means. The only alternative is to dissolve the alloy anodically; this process is the basis for electrochemical machining. Sensors for the detection of both organic and inorganic substances frequently have an electrochemical basis.

It can be argued that the most economically important process having an electrochemical basis is corrosion. A recent study from the National Bureau of Standards [2] indicated that the total cost of corrosion in the US amounts to about 4% of the gross national product, and is now approaching $200 billion per year. Because the basis of corrosion is electrochemical, the basis of countermeasures is frequently electrochemical. Large structures such as storage tanks, ships, and buried pipelines are often protected from corrosion by reducing the potential on the metal in a process known as cathodic protection. The cost of corrosion countermeasures is estimated to be about 10% of the total cost of corrosion, or $20 billion.

Despite the high costs of corrosion, we are fortunate that many metals and alloys spontaneously form protective (passive) films that provide a measure of protection from further corrosion. Aluminum is an example of a metal that would degrade rapidly in air or water were it not for its thin oxide film. By subjecting aluminum to controlled, positive potentials (anodizing), we can improve on the natural film and further protect the metal. In certain environments nickel, zinc, titanium, chromium, iron, and alloys based on these metals also passivate. Without these films, most common metals and alloys could not be used in engineering applications.

1.2 Historical Developments

Seminal discoveries in electrochemistry began around 1800. Crude batteries and electrolytic processes were invented around the turn of the 19th century. Allesandro Volta's "pile" consisted of alternate layers of silver and zinc with a salt-soaked cloth between to carry the current. This crude battery was soon used by Nicholson and Carlisle to decompose water and by Humphry Davy to isolate the active metals sodium and potassium from molten salts. Davy's assistant Michael Faraday made the most important theoretical advances of that era in electrochemistry. The law that bears Faraday's name put the discipline on a firm quantitative basis and helped clarify the differences between quantity of charge and electrical intensity.

In the latter half of the 19th century many technological advances were made. Sir David Grove described a fuel cell in 1839, and Georges Leclanché constructed the carbon-zinc battery in 1868. Aluminum was originally produced by a chemical route, but at $100/lb there were few takers. The Hall-Heroult process of reducing alumina dissolved in cryolite brought the price down to $2/lb soon after its introduction in 1886. About the same time, diaphragm cells for producing chlorine and sodium hydroxide from brine became available and this electrochemical route eventually became the dominant process in the US.

Important theoretical developments in electrochemistry arose from an improved understanding of thermodynamics at the turn of the century. Walther Nernst made several key contributions including the equation named after him. Another significant advance was made by Julius Tafel, who related the potential difference across the double layer to the reaction rate. Theories of the double layer first advanced by Helmholtz were further refined and led to an improved model of electrolyte behavior proposed by Debye and Hückel.

As electrochemical processes became industrially important, increasing effort was devoted to making those processes more efficient. Much of the development in industry was done on a proprietary basis, and serious efforts to quantify the design of electrochemical processes did not evolve until the 1920s. In plating processes the desire to provide a relatively uniform deposit led to test methods for measuring "throwing power" or metal distribution. By the 1940s several investigators recognized that the equations governing the potential in electrolytes had analogies in heat conduction, molecular diffusion, and fluid flow; consequently, the methods of mathematical physics were appropriated for simulation of the behavior of electrochemical cells.

The formal synthesis of electrochemistry and engineering began in the early 1950s with the work of the late Norbert Ibl in Switzerland and Charles Tobias in the United States. Their early contributions included techniques for measuring and quantifying the effects of mass transport, hydrodynamics, electrode kinetics, and gas evolution in electrochemical systems. From these early studies have emerged sophisticated techniques for modeling, simulating, and designing a wide range of electrochemical processes.

Bibliography

[1] *New Horizons in Electrochemical Science and Technology,* Report of the Committee on Electrochemical Aspects of Energy Conservation and Production, National Research Council (Washington, D. C.: National Academy Press, 1986).

[2] L. H. Bennett, J. Kruger, R. L. Parker, E. Passaglia, C. Reiman, A. W. Ruff, H. Yakowitz, and E. B. Berman, *Economic Effects of Metallic Corrosion in the United States,* NBS Special Publication 511-1 (Washington, D. C.: US Government Printing Office, 1978).

5

Chapter 2

Basic Concepts

Because electrochemistry evolved from several disciplines, the conventions relating to cells have, historically, reflected this diversity. Electrode reactions were formerly written as oxidation reactions in the US, but we now follow the IUPAC convention of writing them as reductions. The convention of labeling current as the flow of positive charge derives from early experiments with electricity and has been adopted in electrochemical cells. A consistent set of conventions for electrochemical cells has emerged and will be followed throughout this book. Pertinent constants and conversion factors are listed in Appendix A.

2.1 Conventions

An electrochemical cell consists of at least two electrodes where reactions occur, an electrolyte for conduction of ions, and an external conductor to provide for continuity of the circuit as shown in Fig. 2.1. An elementary thermodynamic analysis of such a cell begins with calculations based on standard electrode potentials. Any measurement of a single electrode must involve a second electrode to complete the circuit; consequently, an electrode potential is always measured with respect to another electrode. The electrode chosen as the universal standard is the hydrogen electrode under specified conditions. The reaction is

$$2H^+ + 2e = H_2 \qquad (2.1)$$

6

At 25°C and unit activity for hydrogen ions, this electrode reaction defines the zero of potential. By connecting another electrode to this half cell and maintaining electrical contact as in Fig. 2.2, one can construct a cell for measuring the potential of the second electrode reaction with respect to the standard hydrogen electrode (SHE). In

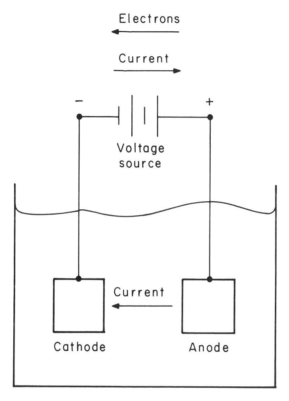

Figure 2.1: Cell schematic. In a driven cell cations migrate toward the cathode, anions toward the anode. Current is defined as the flow of positive charges and moves in a direction opposite to the electrons in the external circuit.

this cell the potential of the electrode on the left (the SHE) is subtracted from the potential of the electrode on the right. This cell potential may be of either sign. Once the signs are established with respect to the SHE, we can always arrange the cell so that a volt-

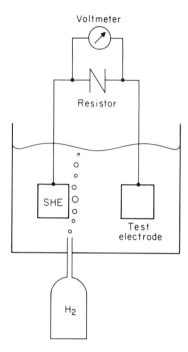

Figure 2.2: Cell for the experimental determination of standard electrode potential. The electrode reaction of interest on the right-hand side is measured with respect to the SHE.

meter gives a positive reading. A list of standard electrode potentials appears in Appendix B. We usually model an electrochemical cell as a device operating at constant temperature and pressure. From thermodynamics we know that the maximum electrical energy available from such a cell is proportional to the change in Gibbs free energy

$$\Delta G^o \;=\; -nFE^o \tag{2.2}$$

where n is the number of electrons participating in the reaction and F is Faraday's constant. The superscript o indicates standard conditions: 25°C and unit activity for the species. A list of the nomenclature used in this book appears before the first chapter. If the Gibbs free energy change is negative for a reaction, then it is thermodynamically favored to proceed in the direction indicated—toward the products. To be consistent with this convention, the standard cell potential must be positive when the products are thermodynamically favored. We will always consider the standard cell potential to be a positive quantity. A consistent method for obtaining positive values is:

1. Choose the two pertinent electrode reactions from a table of standard electrode potentials.

2. Subtract the more negative value from the more positive value. This operation will always result in a positive value for the standard cell potential, provided different reactions take place at the electrodes.

3. Reverse the sense of the more negative reaction, balance the number of electrons, and add the two electrode reactions to obtain the overall reaction.

The resultant potential is an intensive quantity and is unaffected by the number of electrons chosen for an electrode reaction; therefore, in Step 3, we can alter the stoichiometric coefficient of each species (including the electrons) by multiplying through by any positive constant, as long as we eliminate the electrons from the overall cell reaction.

When describing a cell schematically, we place the more positive cell reaction on the right-hand side and the more negative reaction on the left-hand side; phase boundaries are indicated with a solidus. For example, a cell composed of zinc and copper electrodes could be indicated as follows:

$$Zn \ / \ Zn^{2+} \ / \ Cu^{2+} \ / \ Cu$$

Additional information such as solvent type or species concentration is sometimes included. Using the method given above for the calculation of a standard cell potential, we obtain 0.34 V - (-0.76 V) = 1.1 V, which represents the maximum potential obtainable from a single cell under standard conditions.

Electrode Process	Energy Consumer (Electrolytic Cell)	Energy Producer (Galvanic Cell)
Reduction (Cathode)	-	+
Oxidation (Anode)	+	-

Table 2.1: Electrode sign conventions for electrolytic and galvanic cells.

Oxidation occurs at the anode and reduction at the cathode. In the Zn/Cu cell, the zinc metal is oxidized and the copper ions are reduced; thus, the zinc electrode is the anode and the copper electrode is the cathode. In an energy producing (or driving) cell, the zinc electrode is the negative electrode and the copper electrode is the positive electrode. These signs would be obtained if the red lead of a voltmeter were connected to the positive and the black lead to the negative.

If we were to attempt to reverse the flow of current, as we would want to do in a rechargeable battery system, then we would need to put an energy source, such as a power supply, in the external circuit. Under standard conditions our thermodynamic analysis tells us that at least 1.1 V are needed before current reversal can occur. Reversing the current flow implies that the copper metal must be oxidized and the zinc ions must be reduced. Under these conditions, the anode and the cathode must be reversed: the zinc electrode becomes the cathode and the copper electrode becomes the anode. To accomplish this current reversal, we must make the zinc electrode more negative and the copper electrode more positive; therefore, the sign of each electrode remains the same—as a voltmeter would confirm—but the electrochemical reaction changes direction at each electrode. Such a cell is an energy consumer (or driven cell). These conventions are summarized in Table 2.1. We often associate a sign with current depending on whether it is entering or leaving an electrode: positive if it leaves the electrode and negative if it enters the electrode from the electrolyte. In our driving cell, current at the zinc electrode is positive. In general, current is considered positive at the anode and negative at the cathode.

If we are not operating under conditions of standard concentration, we can estimate the thermodynamic electrode (or cell) potential from the Nernst equation

$$E = E^\circ - \frac{RT}{nF} \ln\left(\prod_i c_i^{s_i}\right) \qquad (2.3)$$

where E is the reversible potential of the reaction at the specified concentrations. The convention that states that stoichiometric co-efficients are positive for products implies that products are in the numerator of the fraction in the logarithmic term. The details of this equation are discussed in Ch. 3. Our convention is to subtract the logarithmic term from the standard potential. Some authors put the concentrations of reactants in the numerator and add the logarithmic term to the standard reversible potential; this convention yields the same result as Eq. 2.3.

The shorthand version of an electrode reaction is often written as

$$\sum_i s_i M_i^{z_i} \longrightarrow ne \qquad (2.4)$$

where s_i is the stoichiometric coefficient of species i (with the convention that s_i is positive for products and negative for reactants), M_i is the symbol for the chemical species, and z_i is the charge number of the chemical species.

2.2 Faraday's Laws

The relationship between charge passed and amount of a substance oxidized or reduced at an electrode was first quantified by Michael Faraday in the 1830s. His ideas are embodied in 2 statements pertinent to electrolytic processes.

1) The amount of product formed is directly proportional to the charge passed.

2) For a specified quantity of charge passed, the masses of products formed are proportional to the electrochemical equivalent weights of the products.

These 2 principles can be concisely stated in one equation

$$m = \frac{sMIt}{nF} \tag{2.5}$$

where m is the mass of the substance, s is the stoichiometric coefficient of the species, M is the atomic or molecular weight, I is the current, and t is the time elapsed. An electrochemical equivalent weight is sM/nF. Often, the electrode reaction is written such that the material being produced or consumed has a coefficient equal to one. For example, if we were to calculate the mass of hydrogen consumed in the electrode reaction appearing in Eq. 2.1, s would be equal to one. Because an electrode reaction can always be multiplied by a constant to make the species of interest appear with a coefficient of one, the constant s is often omitted in writing Faraday's law. This equation is valid for a constant current process. In other cases the current must be integrated over time to calculate Q, the total charge passed.

$$Q = \int_0^t I dt \tag{2.6}$$

The standard unit of charge is the coulomb, which is equal to 6.28 $\times$ 10^{18} electrons. The numerical value of this unit of charge arises from Coulomb's law and Maxwell's laws. Coulombs's law relates the force between charged particles to the quantity of charge

$$f = 8.99 \times 10^9 \frac{Q_1 Q_2}{d^2} \tag{2.7}$$

where the force f is in newtons, and the distance between charges d is in meters. The proportionality constant in Coulomb's law is 8.99×10^9 N-m^2/C^2, which is 10^{-7} times the square of the speed of light expressed in m/s; this term comes from relations derived from Maxwell's laws

$$c^2 = \frac{1}{\epsilon_0 \mu_0} = \frac{1}{4\pi\epsilon_0} \frac{4\pi}{\mu_0} \tag{2.8}$$

where c is the speed of light, ϵ_0 is the permittivity of free space $(8.85 \times 10^{-12}$ C^2/N-m$^2)$, and μ_0 is the permeability of free space, $\mu_0/4\pi = 10^{-7} H/m$. The Faraday is the number of charges that must be passed to oxidize or reduce one mole of a compound for a one electron process.

$$\frac{6.02 \times 10^{23} \text{ electrons/equiv}}{6.28 \times 10^{18} \text{ electrons/C}} = 96,500 \frac{\text{C}}{\text{equiv}} \qquad (2.9)$$

Where extreme accuracy is required, a value of 96,485 C/equiv is used. In certain engineering calculations, especially in battery applications, a value of 26.8 ampere hours (A-h) is used for the value of the Faraday. A list of useful conversion factors and derived constants appears in Appendix A.

Faraday's law gives us a theoretical value of the change in mass of a sample for a specified amount of charge passed. There are numerous causes for a deviation from the theoretical value of the change in sample mass, among them: (1) some of the charge is being consumed in parasitic processes; (2) all of the reactants are not consumed; (3) the postulated electrochemical process is not the actual process occurring; or (4) some of the material from the sample falls off (spalls). Under certain circumstances the deviations can give us insight into the nature of the parasitic processes, but more frequently the deviations provide a measure of the efficiency of a process.

2.3   Current and Voltage Efficiency

When we are considering an electrolytic (energy consuming) process, we define a term called the current efficiency, ϵ_c, as

$$\epsilon_c = \frac{\text{actual chemical change (desired)}}{\text{theoretical chemical change}} \qquad (2.10)$$

For example, if we were to plate 20 g of copper from a copper sulfate bath with the passage of 25 A-h, we could determine the current efficiency. The electrode reaction is

$$Cu^{2+} + 2e = Cu$$

From Faraday's law we calculate the theoretical value of the mass of copper plated:

$$m = \frac{MIt}{nF} = \frac{(63.5 \text{ g/mol})(25 \text{ A} - \text{h})}{(2 \text{ mol/equiv})(26.8 \text{ A} - \text{h/equiv})} = 29.6 \text{ g}$$

Our current efficiency is

$$\epsilon_c = \frac{20 \text{ g}}{29.6 \text{ g}} = 0.68$$

This value implies that 68% of the current was effective in reducing copper ions, and the remainder was involved in other processes, possibly the parasitic reduction of hydrogen ions.

In an energy producing (galvanic) process, we will use more reactant than is calculated from Faraday's law, and we define the faradaic efficiency, ϵ_f, as

$$\epsilon_f = \frac{\text{theoretical reactant required}}{\text{amount of reactant consumed}} \tag{2.11}$$

To calculate overall energy efficiency, we need to determine voltage efficiency. In an electrolytic process, we need to supply a larger voltage than is theoretically required, and we define this voltage efficiency, ϵ_v, as

$$\epsilon_v = \frac{\text{theoretical voltage}}{\text{voltage at terminals}} \tag{2.12}$$

In a galvanic process the voltage output is less than the thermodynamic (theoretical) voltage, and the fraction on the right-hand side of the above equation must be inverted. Using these definitions for voltage and current efficiency, we obtain the following energy efficiency ϵ_e from the product of the current and voltage efficiencies:

$$\epsilon_e = \epsilon_c \epsilon_v \tag{2.13}$$

2.4   Ion Conduction

The current flow in an electrolyte results from the movement of positive and negative ions; ion motion is caused by potential and concentration gradients. For the moment we will focus on potential gradients and quantify the resistance offered by the ion conducting medium (electrolyte). Although aqueous electrolytes have been studied intensively, electrolytes can also be in the form of molten salt, polymer, nonaqueous liquid, glass, or supercritical fluid.

Ion conductivity differs fundamentally from electrical conductivity in metals. Typically, the maximum concentration of charged species that can be obtained in an electrolyte is much lower than in a metallic conductor; moreover, the mobility of large (often solvated) ions is much lower than that of electrons. As a consequence, the conductivity of metals is much higher than that of ions in solution. In an aqueous system at room temperature, conductivities are of the order of 10^{-2} ohm^{-1}-cm^{-1} for a 0.1 N salt solution, while the conductivity of a typical metal such as iron, is of the order of 10^5 ohm^{-1}-cm^{-1}. Because the mechanisms of conductivity are different, the temperature dependences of conductivity display different functional relationships. For electrolytes, conductivity increases with increasing temperature:

$$\frac{1}{\kappa} \frac{\partial \kappa}{\partial T} \simeq 2.5\%/^\circ C \qquad (2.14)$$

For the electrical conductivity of metals and alloys, the temperature coefficient of conductivity is negative and about an order of magnitude lower.

The resistance to current flow is characterized by the resistivity ρ, often expressed in ohm-cm. For electrolytes we often use the reciprocal of this quantity, the conductivity, denoted by κ. In general, ion conductivity is proportional to the number of charge carriers and to their mobility; in a few, important special cases the alignment of the charged species also plays a role.

In an aqueous system, the number of charge carriers is proportional to the bulk concentration of an acid, base, or salt only in dilute solution, where all of the chemical species dissociate into ions. In a more concentrated region, the degree of dissociation drops and the solution viscosity increases; thus, both the number and mobility of charge carriers decrease. A plot of conductivity vs. bulk concentration appears in Fig. 2.3. For soluble compounds, conductivity reaches a maximum, but then decreases due to association and viscosity effects. Some salts reach their solubility limits before other effects limit conductivity.

An electrolyte conductivity calculation is usually based on data and corrected for appropriate operating conditions. The temperature dependence shown in Eq. 2.14 can be used to estimate a conductivity

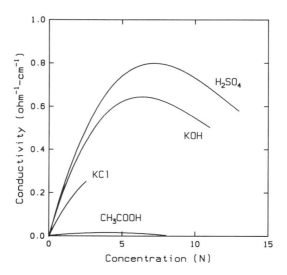

Figure 2.3: Ionic conductivity as a function of concentration at 25°C.

at an operating temperature different from the temperature at which data are available. A calculation based on simple theories requires, at minimum, information on the size of the ionic species and the viscosity of the electrolyte. More comprehensive models require additional information such as degree of dissociation and dielectric constant of the solution.

The most elementary model of ion conduction is based on the assumption of complete dissociation in a continuous medium. For any species i, the ion flux $(mol/cm^2$-s$)$ is

$$\mathbf{N}_i = c_i \mathbf{v}_i \tag{2.15}$$

where $\mathbf{v}_i$ is the velocity of the particle. The current density (A/cm^2) is due to the motion of charged particles.

$$\mathbf{i} = F \sum_i z_i c_i \mathbf{v}_i \tag{2.16}$$

When we substitute Eq. 2.15 into Eq. 2.16, we obtain

$$\mathbf{i} = F \sum_i z_i \mathbf{N}_i \tag{2.17}$$

The ion velocity is proportional to the field strength (V/cm)

$$\mathbf{v}_i \ = \ z_i u_i F \mathbf{E}_f \tag{2.18}$$

where u_i is a proportionality constant called the ionic mobility with units of $\text{cm}^2\text{-mol/J-s}$; note that $F z_i u_i$ has units of $(\text{cm/s})/(\text{V/cm})$, which is the velocity under a unit electric field. Substitution of Eq. 2.18 into Eq. 2.16 yields

$$\mathbf{i} \ = \ F^2 \sum_i z_i^2 u_i c_i \mathbf{E}_f \tag{2.19}$$

The field is proportional to the negative of the potential gradient.

$$\mathbf{E}_f \ = \ -\boldsymbol{\nabla}\phi \tag{2.20}$$

Combining Eqs. 2.19 and 2.20 gives

$$\mathbf{i} \ = \ -\kappa\boldsymbol{\nabla}\phi \tag{2.21}$$

where κ is the ionic conductivity

$$\kappa \ = \ F^2 \sum_i z_i^2 u_i c_i \tag{2.22}$$

Eq. 2.21 is Ohm's law for electrolytic solutions.

One of the models that we can use for ion conduction is that of a charged particle in an electric field. If we assume that ions are spheres in a continuous, viscous medium moving at low values of the Reynolds number, then we can use Stokes's law

$$\mathbf{f} \ = \ 6\pi r \mu \mathbf{v} \tag{2.23}$$

where $\mathbf{f}$ is the drag force on the sphere, r is the radius of the particle, and μ is the viscosity. The dimensionless Reynolds number is

$$Re \ = \ \frac{dv\rho}{\mu} \tag{2.24}$$

where d is the ion diameter and ρ is the density. In a uniform electric field $\mathbf{E}_f$, the force on the charged particle due to the field is

$$\mathbf{f} \ = \ ze\mathbf{E}_f \tag{2.25}$$

where e is the charge on a particle. At constant particle velocity, the net force is zero, and we equate Eq. 2.23 and Eq. 2.25. Solving for the magnitude of the velocity, we obtain

$$v = \frac{zeE_f}{6\pi\mu r} \tag{2.26}$$

For a salt that completely dissociates into a singly charged cation and a singly charged anion (1:1 electrolyte) in an electric field of 1 V/cm, the velocity is

$$v = \frac{(1 \text{ chg})(1.6 \times 10^{-19} \text{C/chg})(1\text{V/cm})(10^7 \text{ erg/J})}{6\pi(10^{-2} \text{ g/cm} - \text{s})(10^{-8}\text{cm})} \tag{2.27}$$

A volt is a J/C and an erg is a g-cm^2/s^2; therefore, the above expression yields a velocity of 8.4×10^{-4} (cm/s)/(V/cm). We would expect this velocity for a 1-Å sphere in unit electric field. For a 1:1 electrolyte, we can calculate the conductivity in an aqueous medium if we can estimate an ionic radius. The crystal radii of lithium, sodium, and potassium are 0.68, 0.95, and 1.33 Å, respectively. For a 0.1 N solution of KCl, the current density from Eq. 2.16 is

$$
\begin{aligned}
i &= F z_+ c_+ \mathbf{v}_+ + F z_- c_- \mathbf{v}_- \\
&= 2 \left(96500 \, \frac{\text{C}}{\text{equiv}} \right) \left(1 \times 10^{-4} \frac{\text{equiv}}{\text{cm}^3} \right) \left(8.4 \times 10^{-4} \frac{\text{cm}}{\text{s}} \right) \tag{2.28} \\
&= 0.016 \, \frac{\text{A}}{\text{cm}^2} \tag{2.29}
\end{aligned}
$$

In this order of magnitude calculation, we have assumed that the anion has same characteristics as the cation.

If we assume that there are no concentration gradients, then we can apply Ohm's law, Eq. 2.21. In a system with a uniform potential gradient, current density is proportional to the potential gradient and is uniform. Comparing Eq. 2.28 with Eq. 2.21, we see that for a unit electric field strength, which was assumed in the velocity calculation, the conductivity is numerically equal to the current density. For KCl, NaCl, and LiCl, the conductivities for 0.1 N aqueous solution at 25°C are 0.0129, 0.0107, and 0.0096 ohm^{-1}-cm^{-1}, respectively. It is surprising that such a simple model can give the correct order of magnitude for the conductivity; however, one of the shortcomings is evident from this comparison with the data. From the model we would expect LiCl to have the highest conductivity of the three salts because lithium has the smallest crystal radius. By neglecting the degree of solvation, we have underestimated the effective radius of the charged particle.

We can now check the assumption that the Reynolds number is small. Substitution of our value of velocity into Eq. 2.24 results in a Reynolds number that is of the order of 10^{-9}, which is sufficiently small to justify the use of Stokes's law.

Several empirical relations are useful in estimating conductivity. In the 1870s Fredrich Kohlrausch carried out an extensive series of investigations to explain and correlate conductivity data. He found it more useful to define a function that does not change abruptly with concentration. The function is called the equivalent conductance, and is expressed by

$$\Lambda = \frac{\kappa}{z_+ \nu_+ c_+} \frac{cm^2}{ohm - equiv} \tag{2.30}$$

where ν_+ is the number of positive ions into which a chemical species dissociates. In Eq. 2.30 we have arbitrarily based the calculation on the positive ions, but because $|z_+ \nu_+| = |z_- \nu_-|$, we could have used the negative ion. For dimensional consistency the concentration must be expressed in mol/cm^3.

Empirically, equivalent conductance was correlated with the square root of concentration. For electrolytes that dissociate to a high degree, a plot of the equivalent conductance vs. square root of concentration yields a nearly linear relation (Fig. 2.4). Because of incomplete dissociation at higher concentrations and higher electrolyte viscosity, equivalent conductance decreases at higher concentrations. An extrapolation to zero concentration gives the equivalent conductance at infinite dilution Λ°. Kohlrausch noted that the difference between Λ° values for pairs of salts having a common ion was approximately constant. For example, the difference in equivalent conductances at infinite dilution, expressed in cm^2/ohm-equiv, for KCl and NaCl is 150 - 128 = 22; for KNO_3 and $NaNO_3$ it is 146 - 123 = 23. Similar results are obtained for pairs of salts with common cations. From these observations, Kohlrausch concluded that the equivalent conductance can be considered to be the sum of two ionic components acting independently.

$$\Lambda^o \ = \ \lambda_+^o \ + \ \lambda_-^o \qquad\qquad (2.31)$$

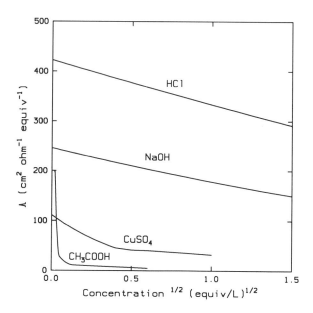

Figure 2.4: Equivalent conductance vs. the square root of concentration at 25°C.

Tables of equivalent conductances appear in Appendix C. We see that most of the equivalent ion conductances at infinite dilution are roughly 50 cm^2/ohm-equiv at 25°C; thus, the equivalent conductances are of the order of 100. Notable exceptions are hydrogen and hydroxide ions, which have ion conductances of approximately 350 and 200 cm^2/ohm-equiv, respectively. An increase of this magnitude cannot be attributed solely to ion size or hydration. The mechanism of ion conduction is altered due to the structure of an electrolyte. Because of the special alignment of ions and solvent, due to interparticle forces, charge can be transferred by a hopping (Grotthuss) mechanism. In aqueous systems forces are usually associated with hydrogen bonding. Recently, Gileadi and co-workers [1] showed that the hopping mechanism also occurs in systems with ions other than hydrogen.

The observation that the product of the equivalent conductance at infinite dilution and the viscosity is approximately constant was due to Walden's observation that

$$\Lambda^{\circ}\mu = \text{constant} \tag{2.32}$$

This relation is approximately correct for a large number of compounds over a wide range of temperatures. It is consistent with the simple model using Stokes's law. According to that model, only the viscosity would be expected to change with temperature. In reality, the degree of solvation—and the effective radius—could also change. The relation does not hold for solutions that have the kind of ordering that leads to the hopping mechanism. In fact, significant deviation from Walden's rule provides evidence of structure in the solution (Fig. 2.5). Mobility, diffusivity, and equivalent ion conductance all pertain to the facility with which an ion moves through a solution. Equivalent ion conductance is related to mobility by

$$\lambda_i = |z_i| F^2 u_i \tag{2.33}$$

In the limit of infinitely dilute solution, the Nernst-Einstein equation relates ionic mobility to diffusivity

$$D_i = RT u_i \tag{2.34}$$

where T is the absolute temperature. Often this expression is used over a range of concentrations when only approximate values are required. A further relation is provided by the Stokes-Einstein equation.

$$\frac{D_i \mu}{T} = \text{constant} \tag{2.35}$$

This equation is essentially a restatement of Walden's rule combined with the Nernst-Einstein equation.

For an electrolyte consisting of one or more ionic species, one can, in principle, estimate the conductivity directly from Eq. 2.22. That equation requires mobility data, which are less frequently tabulated than equivalent ion conductances, but the two quantities are related through Eq. 2.33. Because of interactions among species, Eq. 2.17 will only give reasonable results for dilute solutions. We can see from Fig. 2.6 that because of common ion effects, strong interactions are apparent at higher concentrations of solutes. In fact, increasing the

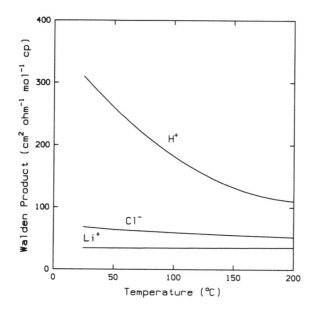

Figure 2.5: Product of viscosity and equivalent ionic conductivity (Walden product) vs. temperature.

copper sulfate concentration leads to reduced conductivity at sulfuric acid concentrations above roughly 0.2 M. This relationship is contrary to the dilute solution predictions of Eq. 2.22, where increasing ion concentration is expected to lead to increased conductivity. For making estimates of conductivity in the absence of data, a large number of semi-theoretical correlations are available. Formulas for estimating conductivity as a function of concentration or temperature are presented by Horvath [3].

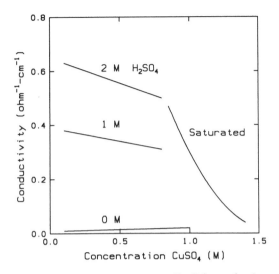

Figure 2.6: Conductivity of aqueous $CuSO_4$ solutions with H_2SO_4 added as a supporting electrolyte at 25°C.

2.5 Transference Numbers

We know from Eq. 2.16 that current arises from the motion of positively and negatively charged particles. Because the mobilities of the positive and negative species are never exactly equal, there will be a tendency for a species of higher mobility to migrate more rapidly under the influence of an electric field, as predicted by Eq. 2.18. To preserve electrical neutrality in a solution, the net amount of charge passed at each electrode must be equal and opposite; however, this constraint does not imply that the charge carried by the various species in solution must be borne equally by the positive and negative charge carriers. From Eq. 2.16 we recognize that the amount of current carried by a species is proportional to its concentration. Consequently, we expect that most of the current will be carried by mobile species in high concentration.

The transference number t_j (sometimes called the transport number) represents the fraction of current carried by a specified ion in the absence of concentration gradients. The fraction of current carried by species j is given by Eq. 2.21

$$t_j \;=\; \frac{i_j}{i} \tag{2.36}$$

$$=\; \frac{F^2 z_j^2 u_j c_j \boldsymbol{\nabla} \phi}{F^2 \displaystyle\sum_i z_i^2 u_i c_i \boldsymbol{\nabla} \phi} \tag{2.37}$$

$$=\; \frac{z_j^2 u_j c_j}{\displaystyle\sum_i z_i^2 u_i c_i} \tag{2.38}$$

The fractional current carried by each species must add up to the total current

$$\sum_i t_i \;=\; 1 \tag{2.39}$$

Equation 2.38 can be simplified for a binary electrolyte. Because the solution is electrically neutral, we have

$$|z_+ c_+| \;=\; |z_- c_-| \tag{2.40}$$

When we substitute Eq. 2.40 and Eq. 2.33 into Eq. 2.38, we obtain for a binary electrolyte

$$t_+ \;=\; \frac{\lambda_+}{\lambda_+ + \lambda_-} \tag{2.41}$$

$$=\; \frac{\lambda_+}{\Lambda} \tag{2.42}$$

When concentration gradients are present, some of the current arises from diffusion, but the expression for the transference number only accounts for the fraction of current due to migration. Although the above expressions are generally valid, the transference number no longer exactly represents the fraction of current carried by a species when concentration gradients are present. The transference number is generally a weak function of concentration and temperature. Values of transference numbers for common ionic species appear in Table 2.2. If we consider a binary electrolyte such as $CuSO_4$ in water, the primary current-carrying species in the dilute solution are cupric and sulfate ions. The transference number for cupric ions is 0.44. Because the

Electrolyte	Concentration (N)	Temperature (°C)	t_+
KCl	*dil.*	25	0.49
KCl	*dil.*	45	0.49
KCl	1.0	25	0.49
HCl	*dil.*	25	0.82
HCl	*dil.*	45	0.80
HCl	0.1	25	0.83
NaCl	0.1	25	0.39
CaCl$_2$	0.1	25	0.41

Table 2.2: Cation transference numbers as a function of concentration and temperature. Infinite dilution is indicated by *dil.*

sum of the transference numbers must equal one, the transference number for the sulfate ion is 0.56. Typically, salts in aqueous solution have transference numbers near 0.5, and each species carries roughly half the current. For acids, the cation transference number is much higher, reflecting the greater mobility of the hydrogen ion.

To increase solution conductivity, an additional component that does not participate in the electrode reactions is often added to a salt solution. The non-reactive component is called a supporting electrolyte. Its use is illustrated in Fig. 2.6. Because the solubility and mobility of cupric and sulfate ions are relatively low, the addition of sulfuric acid significantly increases solution conductivity. The presence of a supporting electrolyte also changes the transference number of the copper and sulfate ions. If we add the supporting electrolyte in great excess, both the concentration and mobility of the hydrogen ion tend to increase its transference number, as Eq. 2.38 indicates.

Consider, for example, an electrolytic cell with two copper electrodes, CuSO$_4$ solution, and an excess of sulfuric acid. Although the transference number for the cupric ion is low, we cannot conclude that the cupric ion carries essentially no current. At steady state, cupric ions must be physically transported across solution from the anode to the cathode. We infer that little current is carried by the cupric ion through migration. Because concentration gradients must be present at the electrodes, diffusion plays a role in the transport. In an excess supporting electrolyte the role of diffusion is magnified because, at constant current, the field is reduced in the higher conductivity medium.

2.6   Problems

1. An important industrial reaction is the electrolysis of brine to form chlorine and sodium hydroxide. At one of the electrodes chloride ions are oxidized to form chlorine. The overall reaction is

$$2NaCl + 2H_2O \rightarrow 2NaOH + Cl_2 + H_2$$

Determine the electrode reactions. Identify the anode, positive electrode, and direction of current flow. Calculate the standard cell potential.

2. High purity hydrogen is produced commercially by the electrolysis of water in basic solution (5 M KOH).

$$2H_2O \rightarrow 2H_2 + O_2$$

a) Identify the electrode reactions. Note that in 5 M KOH, the concentration of hydrogen ions is extremely low, and the reaction in Eq. 2.1 is not involved.
b) Calculate the standard cell potential.
c) Calculate the minimum amount of charge required to produce 1 kg of hydrogen.
d) Estimate the conductivity of pure water and compare this value with the conductivity of 5 M KOH at room temperature.
e) The industrial cell for electrolyzing water is operated at 80°C. Estimate the conductivity of the electrolyte at this temperature from Eq. 2.14. The actual conductivity is 1.4 ohm^{-1}-cm^{-1}. Why is the discrepancy between the two values so large in this case?

3. Calculate the production rate of aluminum in a 100,000 A reduction cell at 95% current efficiency. The overall reaction is

$$2Al_2O_3 + 3C \rightarrow 4Al + 3CO_2$$

4. Most fuel cells presently being developed use hydrogen for the fuel. Hydrogen is oxidized at the anode and oxygen is reduced at the cathode. The overall reaction is the reverse of the water electrolysis reaction. The faradaic efficiency in a test cell is 90% and the operating voltage is 0.7 V. Estimate the rate of hydrogen consumption (kg/day) for a 1 kW cell.

5. Use Eqs. 2.22 and 2.33 to estimate the conductivity of a solution that is 0.5 M $CuSO_4$ and 1 M H_2SO_4. Compare the value with that from Fig. 2.6.

6. Estimate the concentration of H_2SO_4 required to reduce the transference number of copper ions in a 0.05 M $CuSO_4$ to 0.1.

Bibliography

[1] I. Rubinstein, M. Bixon, and E. Gileadi, *J. Phys. Chem.*, **84,** 715 (1980).

[2] E. F. Kern and M. Y. Chang, *Trans. Am. Electrochem. Soc.*, **41,** 181 (1923).

[3] A. L. Horvath, *Handbook of Aqueous Electrolyte Solutions: Physical Properties, Estimation, and Correlation Methods* (New York: Halstead Press, 1985).

Chapter 3

Thermodynamics

From a thermodynamic analysis of an electrochemical cell, we can determine the direction and magnitude of the driving forces for potential electrochemical reactions. A negative result gives definite information. A reaction that is not thermodynamically favored does not proceed to a significant degree and need not be further analyzed. In a sense a positive result gives less information. A thermodynamically favored reaction does not necessarily proceed at a significant rate. To perform an analysis on a cell, additional knowledge of system kinetics and mass transport conditions is required. Electrochemical cells may be limited by high activation energy barriers, which severely limit the overall reaction rate. The common techniques for overcoming kinetic restrictions in electrochemical systems are similar to techniques used in ordinary chemical reactions—providing effective catalysis and increasing operating temperatures. Mass transport limitations are generally overcome by increasing both electrolyte agitation and the concentration of reacting species.

3.1   Cell Thermodynamics

To carry out a thermodynamic analysis, we need to relate thermodynamic (reversible) potential to state variables. Usually, we treat an electrochemical cell as a system operating at constant temperature and pressure. If we consider a closed system operating under these conditions, then we can carry out a thermodynamic analysis. In ther-

29

modynamic terms a closed system is defined as one where transport of material between the system and surroundings is not permitted. From the first law of thermodynamics, the total change in internal energy ΔU of a closed system is equal to the heat added to the system q minus the work done by the system w:

$$\Delta U = q - w \tag{3.1}$$

The work can be divided into work associated with mechanical changes and work associated with other sources, e.g., magnetic, surface, or electrical work. In our analysis it is only the latter form of work that we consider. In a reversible system at constant temperature and pressure, the mechanical contribution due to volume changes is

$$w_p = P\Delta V \tag{3.2}$$

and the total work done by the system is

$$w = w_p + w_e \tag{3.3}$$

where w_e is the electrical contribution to the work. For a reversible change at constant temperature, the heat transferred is given by

$$q = T\Delta S \tag{3.4}$$

where ΔS is the change in entropy.

The natural (or cannonical) state variable for a system operating at constant temperature and pressure is the change in Gibbs free energy, ΔG:

$$\Delta G = \Delta H - T\Delta S \tag{3.5}$$

and the enthalpy change is defined by

$$\Delta H = \Delta U + P\Delta V \tag{3.6}$$

Combining Eqs. 3.1 to 3.4, we have

$$\Delta U = T\Delta S - P\Delta V - w_e \tag{3.7}$$

Substitution of Eqs. 3.6 and 3.7 into 3.5 yields

$$\Delta G = -w_e \tag{3.8}$$

Eq. 3.8 tells us that the electrical work we obtain from a closed system at constant temperature and pressure, operating reversibly, is equal to the change in Gibbs free energy. Because we considered a reversible system, the work that we calculated was the maximum electrical work obtainable from the system; any irreversible process will yield less electrical energy.

The maximum electrical energy available in an external circuit is equal to the number of charges multiplied by the maximum potential difference, which is the reversible cell potential. For one mole of a reactant, the total number of charges is equal to the number of charges participating in the reaction multiplied by Faraday's constant, and the maximum electrical work is given by

$$w_e = nFE \tag{3.9}$$

Equating expressions 3.8 and 3.9, we have

$$\Delta G = -nFE \tag{3.10}$$

In Ch. 2 we defined reversible potential as a positive quantity corresponding to a negative Gibbs free energy change. Equation 3.10 is the key equation relating electrical potential to the traditional thermodynamic framework. We can calculate the free energy of reaction from free energy of formation $\Delta G_{i,f}$ data by

$$\Delta G = \sum_i s_i \Delta G_{i,f} \tag{3.11}$$

where s_i is the stoichiometric coefficient (positive for products and negative for reactants). With Eq. 3.10 and others derived from it, we can relate thermal data to electrical measurements. An abbreviated table of standard electrode potentials appears in Table 3.1; a more comprehensive list is tabulated in Appendix B. A table of electrode potentials gives us information regarding the direction of the thermodynamic driving force. If we hold an electrode at a potential above the reversible potential, oxidation is favored; with the conventions used in Table 3.1 this result implies that the reactant side is favored. If we hold the electrode at a potential more negative than the reversible potential, then reduction is favored, and the reaction is driven to the right.

Electrode Reaction	Standard Electrode Potential (V)
$Au^{3+} + 3e = Au$	1.52
$Cl_2 + 2e = 2Cl^-$	1.36
$O_2 + 4H^+ + 4e = 2H_2O$	1.23
$Ag^+ + e = Ag$	0.80
$Cu^+ + e = Cu$	0.52
$O_2 + 2H_2O + 4e = 4OH^-$	0.40
$Cu^{2+} + 2e = Cu$	0.34
$2H^+ + 2e = H_2$	0.00
$Fe^{2+} + 2e = Fe$	-0.44
$Zn^{2+} + 2e = Zn$	-0.76
$Al^{3+} + 3e = Al$	-1.66
$Li^+ + e = Li$	-3.05

Table 3.1: Standard electrode potentials at 25°C and 1 atm.

We can calculate the standard reversible potential of a battery comprised of a zinc anode and copper cathode. From Table 3.1 we calculate a value of E^o = 0.34 V - (-0.76 V) = 1.1 V. If we make a measurement of the potential under standard conditions, we should obtain approximately the same value. Such a measurement would need to be made with a high impedance voltmeter so that the amount of current drawn would be small, and the indicated potential at the terminals would be close to the reversible potential. The maximum electrical work available from the Zn-Cu battery is given by

$$
\begin{aligned}
w_e &= nFE \\
&= (2 \text{ equiv})(96,500 \text{ C/equiv})(1.1 \text{ V})[(1 \text{ (J/C)})/\text{V}] \\
&= 212,000 \text{ J}
\end{aligned}
$$

The Gibbs free energy change for the reaction is $\Delta G^o = - w_e$. This example illustrates that one can obtain values of thermodynamic state functions from strictly electrochemical measurements or calculations.

Because the Gibbs free energy change is a state function, we can calculate the reversible potential of any reaction that can be derived from the algebraic addition of two or more reactions for which pertinent thermodynamic data are available. For example, we can calculate the standard reversible potential of the reaction

$$Cu^{2+} + e = Cu^+ \qquad (3.12)$$

from the reactions

$$Cu^{2+} + 2e = Cu \qquad (3.13)$$
$$Cu^+ + e = Cu \qquad (3.14)$$

If we treat these two reactions as algebraic expressions and subtract Eq. 3.14 from Eq. 3.13, we obtain Eq. 3.12. We can treat the free energy changes of the reactions in the same algebraic manner. Designating Eq. 3.13 as equation 1, Eq. 3.14 as equation 2, and Eq. 3.12 as equation 3, we have

$$\Delta G_1^o - \Delta G_2^o = \Delta G_3^o$$
$$n_1 F E_1^o - n_2 F E_2^o = n_3 F E_3^o \qquad (3.15)$$

Values of the standard reversible potentials for equations 1 and 2 are in Table 3.1. Dividing Eq. 3.15 by F and substituting reversible potential values into Eq. 3.15, we obtain

$$(2)(0.34) - (1)(0.52) = (1)(E_3^o)$$
$$E_3^o = 0.16 \text{ V}$$

3.2 Temperature and Pressure Effects

Table 3.1 gives electrode potentials at standard temperature. We can calculate the reversible electrode potential at other temperatures by calculating the Gibbs free energy at a specified temperature and using Eq. 3.10 to relate the reversible potential to the Gibbs free energy. If we consider a reversible process where only mechanical work is permitted, then we can state the first law as

$$dU = TdS - PdV \qquad (3.16)$$

The enthalpy is defined as

$$H = U + PV \qquad (3.17)$$

and a differential change in enthalpy is expressed by

$$dH \;=\; dU \;+\; PdV \;+\; VdP \tag{3.18}$$

The Gibbs free energy is defined by

$$G \;=\; H \;-\; TS \tag{3.19}$$

and a differential change in free energy is

$$dG \;=\; dH \;-\; TdS \;-\; SdT \tag{3.20}$$

Combining Eqs. 3.16, 3.18, and 3.20, we have

$$dG \;=\; VdP \;-\; SdT \tag{3.21}$$

If we consider a process proceeding from state 1 to state 2, we can write Eq. 3.21 for each state and define $\Delta G = G_2 - G_1$; at constant pressure Eq. 3.21 becomes

$$\left(\frac{\partial \Delta G}{\partial T}\right)_P \;=\; -\Delta S \tag{3.22}$$

Substitution of Eq. 3.10 into Eq. 3.22 yields

$$\left(\frac{\partial E}{\partial T}\right)_P \;=\; \frac{\Delta S}{nF} \tag{3.23}$$

We are primarily interested in systems where changes in thermodynamic functions are due to electrode reactions. In most applications it is the change in a thermodynamic function for the overall reaction that we use; a ΔG term, for example, represents the free energy change for the reaction. Because Eq. 3.22 involves only state functions, the details of carrying out a particular reaction are unimportant. Although Eq. 3.22 was derived under more restricted conditions, it must be generally true for any reversible process carried out in a closed system at constant pressure.

From Eq. 3.23 we can calculate reversible potential over a range of temperatures. Over a small temperature range the assumption of constant entropy change of reaction is usually justifiable, and the integration of Eq. 3.23 yields

$$E_2 - E_1 = \frac{\Delta S}{nF} (T_2 - T_1) \tag{3.24}$$

If the entropy change varies significantly with temperature, then an integration of the temperature-dependent entropy function in Eq. 3.23 would be necessary.

When gaseous products or reactants are involved, much of the contribution to the entropy term arises from translational energy of the molecule; consequently, the sign of the entropy change is usually the same as the change in the number of moles of gaseous components (ΔN). For example, in a fuel cell using hydrogen as the fuel, we have the overall reaction

$$H_2 + \frac{1}{2} O_2 = H_2O \tag{3.25}$$

At room temperature, water is in the liquid state, and the change in the number of moles of gaseous components is $0 - 1 - 1/2 = -3/2$. Using Eq. 3.5, we obtain

$$\Delta S = \frac{\Delta H - \Delta G}{T}$$
$$\Delta S = -163 \text{ J/K}$$

This equation is useful because enthalpy and free energy values are more frequently tabulated in standard references than entropy values. The magnitude of the change in reversible potential with temperature, calculated above, is typical for a large number of reactions (See Table 3.2). It is important to specify the state of each component when making a calculation; note that the change in the number of gaseous moles (ΔN) changes for Reaction 3.25 at higher temperatures where water is in a gaseous state. The change in the reversible potential with pressure can also be calculated from Eq. 3.21:

$$\left(\frac{\partial \Delta G}{\partial P} \right)_T = \Delta V \tag{3.26}$$

Substitution of Eq. 3.10 into Eq. 3.26 gives

$$\left(\frac{\partial E}{\partial P} \right)_T = \frac{-\Delta V}{nF} \tag{3.27}$$

Reaction	E° (V)	ΔG° (kJ)	ΔH° (kJ)	$\partial E/\partial T$ (mV/K)
$H_2 + \frac{1}{2}O_2 \rightarrow H_2O$	1.23	-237	-286	-0.8
$C_2H_6 + \frac{7}{2}O_2 \rightarrow 2CO_2 + 3H_2O$	1.09	-1470	-1560	-0.2
$NH_3 + \frac{3}{4}O_2 \rightarrow \frac{1}{2}N_2 + \frac{3}{2}H_2O$	1.23	-356	-383	-0.3
$Pb + 2HgCl \rightarrow PbCl_2 + 2Hg$	0.54	-103	-95	+0.1
$CuSO_4 + Zn \rightarrow ZnSO_4 + Cu$	1.10	-212	-234	-0.4

Table 3.2: Temperature variation of standard cell potentials.

If the ideal gas assumption is valid for the gaseous components, then we can evaluate Eq. 3.27 in a straightforward fashion.

$$\Delta V = \frac{\Delta N R T}{P} \qquad (3.28)$$

Substitution of Eq. 3.28 into Eq. 3.27 gives

$$dE = -\frac{\Delta N R T}{n F} d\ln P \qquad (3.29)$$

Integrating yields

$$E_2 - E_1 = -\frac{2.303 \,\Delta N R T}{n F} \log\left(\frac{P_2}{P_1}\right) \qquad (3.30)$$

For a one electron process, $2.303 \, RT/nF$ is approximately 60 mV. Because the magnitude of ΔN is on the order of one, the change in reversible potential with pressure is on the order of 10 mV/decade; for example, a change of pressure from one atmosphere to ten atmospheres causes a change in reversible potential of a few tens of millivolts. When only condensed phases (solids and liquids) are involved, changes in reversible potential with pressure are negligible.

Enthalpy change values can be obtained from electrochemical measurements through the Gibbs-Helmholtz equation. Substituting Eq. 3.23 and Eq. 3.10 into Eq. 3.5, we have

$$-nFE = \Delta H - nFT \left(\frac{\partial E}{\partial T}\right)_P \qquad (3.31)$$

or

$$\Delta H = -nFE + nFT \left(\frac{\partial E}{\partial T}\right)_P \qquad (3.32)$$

An evaluation of the enthalpy change through Eq. 3.32 can be made from a series of reversible potential measurements as a function of temperature. For reactions where electrochemical measurements are practical, Eq. 3.32 gives accuracy comparable to thermal determinations. A number of other thermodynamic properties can be obtained from electrochemical measurements. For example, the entropy change of a reaction can be obtained directly from Eq. 3.23.

3.3   The Nernst Equation

To calculate changes in reversible potential when we are not operating at standard conditions, we consider the following thermodynamic relationship:

$$\Delta G = \Delta G^\circ + RT \ln \left(\prod a_i^{s_i}\right) \qquad (3.33)$$

where a_i is the activity of species i. Substituting Eq. 3.10 into Eq. 3.33, we have

$$E = E^\circ - \frac{RT}{nF} \ln \left(\prod a_i^{s_i}\right) \qquad (3.34)$$

As an approximation, we can ignore activity coefficient corrections and use concentrations in place of species activities, and we obtain the Nernst equation, which is

$$E = E^\circ - \frac{RT}{nF} \ln \left(\prod c_i^{s_i}\right) \qquad (3.35)$$

There are several points to note when applying this equation. First, we are neglecting both activity coefficients and the potential that arise from bringing two different liquid phases into contact (liquid-junction potential); this latter potential arises from differing mobilities of ions from the two different phases and is discussed in Ch. 4. Second, this equation cannot be applied to make both temperature and concentration corrections simultaneously. To make temperature corrections, one must first apply Eq. 3.23 to the standard potential and then apply the Nernst equation (at the new temperature). Third, a number of different conventions are commonly applied to the definition of the standard state when using the activity. For substances in excess (e.g., water in a dilute aqueous system) or for solids, the concentration is taken to be one, corresponding to unit activity. These conventions are discussed in greater detail in Ch. 4.

3.4   The Pourbaix Diagram

One common method of representing the Nernst equation is to plot reversible potential vs. pH. This idea was proposed by Marcel Pourbaix [1], who catalogued the behavior of many electrochemical systems. This type of plot is commonly used in corrosion and extractive metallurgical applications. In many cases we know the nature of species in solution or the crystal structure of a solid phase; in some cases the postulated reactions represent our best estimate of the species involved. If we consider the lead-water system, we can determine the phase in equilibrium with water at different pH values from thermodynamic information. For reference, the lines representing hydrogen and oxygen evolution are plotted. In acid solution the Nernst equation for the hydrogen evolution reaction

$$2H^+ \; + \; 2e \; = \; H_2 \tag{3.36}$$

is

$$E \; = \; E^o \; - \; \frac{2.303 \, RT}{nF} \; \log\left(\frac{P_{H_2}}{c_{H^+}^2}\right) \tag{3.37}$$

At room temperature RT/F is 26 mV. If we define the standard state for hydrogen to be gas at one atmosphere, then the numerator of the logarithmic term is unity. The pH is the negative logarithm of the hydrogen ion concentration, and Eq. 3.37 can be written as

$$E \; = \; 0.00 - 0.059 \, pH \tag{3.38}$$

On the Pourbaix diagram (Fig. 3.1), the line representing Eq. 3.38 is diagonal (line a) with a slope of 59 mV per pH unit, i.e., the reversible potential of the hydrogen reaction is reduced by 59 mV for each factor of ten change in hydrogen ion concentration. The labels on the lines

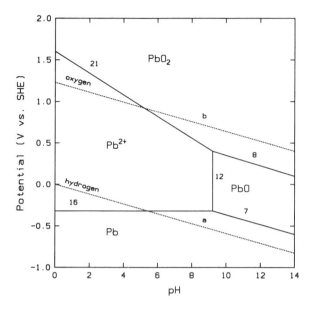

Figure 3.1: Pourbaix diagram for the lead-water system at 25°C

correspond to those used on the Pourbaix diagram in the original reference [1]. Similarly, the oxygen evolution reaction is

$$\frac{1}{2}O_2 + 2H^+ + 2e = H_2O \qquad (3.39)$$

We take the concentration of water to be one, corresponding to unit activity, and the Nernst equation to be

$$E = 1.23 - \frac{2.303\ RT}{nF}\ \log\left(\frac{c_{H_2O}}{P_{O_2}^{0.5}c_{H^+}^2}\right)$$
$$E = 1.23 - 0.059\ \text{pH} \qquad (3.40)$$

From the plot of Eq. 3.40, we see that the oxygen evolution line (line *b*) is parallel to the hydrogen evolution line (line *a*), with the 1.23 V difference maintained at all pH values. We know that the overall reaction between hydrogen and oxygen does not involve hydrogen ions (or hydroxide ions), and consequently we do not expect a pH dependence for the reversible potential of the cell reaction.

The reaction for lead ions being reduced to lead is

$$Pb^{2+} + 2e = Pb \tag{3.41}$$

and the Nernst equation can be expressed as

$$E = E^{\circ} - \frac{2.303\ RT}{nF} \log\left(\frac{1}{[Pb^{2+}]}\right)$$

$$E = -0.126 - 0.0295\ p[Pb^{2+}] \tag{3.42}$$

where the p notation represents the negative logarithm of a quantity and the brackets indicate concentration in molar units. On a detailed Pourbaix diagram, lines representing several values of lead ion concentration often appear. By convention the concentration of ions in solution is frequently assigned a value of 10^{-6} M. Note that Eq. 3.42 is independent of pH and consequently appears as a horizontal line (line 16) on the diagram.

A reaction not involving electron transfer is

$$Pb^{2+} + H_2O = PbO + 2H^+ \tag{3.43}$$

The equilibrium constant is

$$K = \frac{[PbO][H^+]^2}{[Pb^{2+}][H_2O]} \tag{3.44}$$

Because lead oxide is a solid and water is in excess, we consider their concentrations to be unity. The logarithm of the equilibrium constant from standard references is 12.7. The relation between lead ion concentration and pH is

$$K\ [Pb^{2+}] = [H^+]^2$$

$$pK + p[Pb^{2+}] = 2pH$$

$$-12.7 = p[Pb^{2+}] - 2pH$$

$$pH = 6.35 + \frac{1}{2}\ p[Pb^{2+}] \tag{3.45}$$

Because Eq. 3.43 does not involve electron transfer, it is not sensitive to potential and is represented by a vertical line (line 12) on the diagram.

Three additional reactions are

$$Pb + H_2O = PbO + 2H^+ + 2e \qquad (3.46)$$

$$Pb^{2+} + 2H_2O = PbO_2 + 4H^+ + 2e \qquad (3.47)$$

$$3PbO + H_2O = Pb_3O_4 + 2H^+ + 2e \qquad (3.48)$$

The Nernst equations corresponding to the above reactions are

$$E = 0.25 - 0.059 \text{ pH} \qquad (3.49)$$

$$E = 1.45 + 0.029 \text{ p}[Pb^{2+}] - 0.118 \text{ pH} \qquad (3.50)$$

$$E = 0.97 - 0.059 \text{ pH} \qquad (3.51)$$

These three equations are represented as lines 7, 21, and 8 on our simplified Pourbaix diagram.

From the diagram we can make several observations based solely on thermodynamic arguments. We might expect lead to corrode in low pH solution at potentials between roughly 0 and 1 vs. SHE; in that region the soluble lead ion is favored. If we lower the potential below -0.4 V in acid solutions, the solid metal is thermodynamically favored, and we expect to minimize lead corrosion. Reducing the corrosion rate by lowering potential is the basis of cathodic protection. By lowering the potential to -0.6 V in acid solution, we are below the potential where the reduction of hydrogen ions is favored (line *a*), and we expect the evolution of hydrogen gas to occur. By raising the potential in acid medium to 2 V, we encounter another stable phase (PbO_2). Such a phase would be expected to protect the underlying metal and to reduce the corrosion rate.

Most of these predictions, based on thermodynamic arguments, are realized for this system. Because phenomena that are not thermodynamic—kinetics or mass transport—may be rate-limiting, Pourbaix diagrams must be used with caution. For example, the kinetics of film formation may be slow or metastable phases may be formed. Even the range of solvent stability varies widely with the composition of the electrode, and the potential at which hydrogen evolves from water cannot be reliably predicted from a Pourbaix diagram.

In summary, the Pourbaix diagram gives an estimate of the equilibrium composition of an electrode-electrolyte system. The assumptions involved in the Nernst equation are used in all calculations. The orientation of the line on the diagram is dependent on the process: vertical lines represent reactions with no electron transfer, horizontal lines represent reactions with no pH dependence, oblique lines represent reactions with both electron transfer and pH dependence.

3.5   The Equilibrium Constant

The Nernst equation relates concentration changes to changes in reversible potential. If we measure cell potential with a high impedance voltmeter, very little reaction occurs and we make the assumption that the phases are essentially in equilibrium. This equilibrium state will remain undisturbed as long as we do not draw significant current through the external circuit. Upon closing the circuit, the original equilibrium will be disturbed and reaction will continue until the driving force for the reactions is reduced to zero. It is clear from the Nernst equation that a set of concentrations exists that will result in a reversible potential of zero. This state corresponds to the point at which the driving force for further electrochemical reaction is zero, and it represents a stable equilibrium state for the cell. Ratios of concentrations—or more accurately, activities—causing the reversible potential to be equal to zero are used in the definition of the thermodynamic equilibrium constant K.

$$\begin{aligned} E &= E^{\circ} - \frac{RT}{nF} \ln K \\ &= 0 \end{aligned}$$

Therefore,

$$E^{\circ} = \frac{RT}{nF} \ln K \tag{3.52}$$

From a standard cell potential we can immediately obtain an equilibrium constant and determine the extent to which a reaction can theoretically proceed. As an illustration, we can calculate the equilibrium constant for the Daniell cell, which is

$$\text{Zn} / \text{Zn}^{2+} / \text{Cu}^{2+} / \text{Cu} \tag{3.53}$$

The standard cell potential for this reaction is 1.1 V, and the equilibrium constant can be calculated from Eq. 3.52.

$$\ln K = \frac{nFE^\circ}{RT}$$
$$= (76.9)(1.1)$$
$$= 84.6$$
$$\frac{[Zn^{2+}]}{[Cu^{2+}]} = 6 \times 10^{36}$$

The equilibrium constant is a large number, indicating that the reaction can proceed essentially to completion in the absence of impediments.

3.6   Reversible Heat Transfer

If we consider our model electrochemical cell operating at constant temperature and pressure, we expect to have heat transferred during operation. Some of the heat originates from irreversibilities such as ohmic heating and overpotentials. Because the free energy change of a reaction is rarely the same as the enthalpy change of the reaction, we also have an entropy change of reaction, which is given by Eq. 3.5.

$$\Delta S = \frac{\Delta H - \Delta G}{T} \tag{3.54}$$

The $T\Delta S$ term is usually less than 20% of the heat of reaction. Prior to clarifications by Gibbs in the late nineteenth century, it was thought that ΔG and ΔH were equivalent in electrochemical cells.

From Eq. 3.4 we know that there is a reversible heat transfer proportional to the entropy change.

$$q = T\Delta S \tag{3.55}$$

This relationship implies that even for a cell operating reversibly, the cell will transfer heat to (or from) its surroundings. From our convention that heat added is a positive quantity, we see that a positive entropy change of reaction corresponds to heat absorbed by the cell. Generally, for reactions of electrochemical interest, the enthalpy change is a larger negative number than the Gibbs free energy change (Table 3.2); this difference corresponds to a negative entropy change, which implies that heat is released from the cell. For a few electrochemical reactions, the entropy change is positive, and heat is absorbed from the surroundings during reversible operation.

Total heat transfer depends on the sum of the reversible and irreversible losses. The former may be positive or negative corresponding to heat gain or loss, whereas the latter are always negative corresponding to heat loss. For most industrial processes operated near room temperature, the emphasis on high rate leads to high rates of heat production; consequently, irreversible losses predominate and heat loss from the cell results.

Although we have only considered closed systems so far, the treatment of open systems is similar [2]. If we neglect changes in kinetic and potential energies, the isothermal steady-state energy and entropy balances are

$$\sum_i (m_i \overline{H}_i)_{out} - \sum_i (m_i \overline{H}_i)_{in} = \dot{q} - \dot{w}_e \qquad (3.56)$$

$$\sum_i (m_i \overline{S}_i)_{out} - \sum_i (m_i \overline{S}_i)_{in} = \frac{\dot{q}}{T} - \dot{S}_{gen} \qquad (3.57)$$

The overbar represents a partial molar quantity, which is the differential change in a property when a differential amount of substance i is added to a mixture at constant temperature and pressure. The dot over the symbol represents a rate, and m_i is a molar flow rate. For example, the partial molar enthalpy is defined by

$$\overline{H}_i = \left(\frac{\partial H}{\partial n_i} \right)_{T,p,n_j,j \neq i} \qquad (3.58)$$

If we eliminate $\dot{q}$ between these equations, we obtain

$$-\dot{w}_e = \sum_i (m_i \overline{H}_i - T\overline{S})_{out} - \sum_i (m_i \overline{H}_i - T\overline{S})_{in} + T\dot{S}_{gen} \quad (3.59)$$

$$-\dot{w}_e = \sum_i (m_i \overline{G}_i)_{out} - \sum_i (m_i \overline{G}_i)_{in} + T\dot{S}_{gen} \qquad (3.60)$$

For a reversible process there is no entropy production, and the last term in Eq. 3.60 is zero. If the reactants do not mix upon entering the electrochemical device, then the free energy of mixing terms are also zero. For example, consider a flow device such as a fuel cell operating reversibly. If the reactants are hydrogen at the anode and oxygen at the cathode fed in stoichiometric proportions, then for each mole of hydrogen per unit time fed to the device, the maximum work obtainable can be determined from Eq. 3.60. For pure component reactants, the partial molar free energy is just the free energy per mole. If we consider a unit time period, then Eq. 3.60 reduces to Eq. 3.8. In this particular case where there was no mixing of the reactants, an analysis of a closed or open system yielded the same result. For the constant temperature and pressure process, it is the Gibbs free energy change that is directly related to the available electrical energy. In cases where reactants are mixed, the partial molar quantities would no longer be equal to the pure component quantities, and a rigorous calculation would require the use of Eq. 3.60. The free energy of mixing is generally small compared to the free energy of reaction; therefore, in most systems a good estimate could be made from pure component values.

3.7 Problems

1. Methane is being considered for use as the fuel in a fuel cell. The overall reaction is

$$CH_4 + 2O_2 \rightarrow CO_2 + 2H_2O$$

In acid electrolyte, oxygen is reduced at the cathode.

a) Write the electrode reactions. Note that a balanced reaction can be written using methane and oxygen as anodic reactants and oxygen as the cathodic reactant. Try it! Why is this scheme unreasonable?

b) Estimate the standard cell potential.

c) Calculate the change in reversible potential with temperature (mV/K) near room temperature.

d) Calculate the change in reversible potential with pressure (mV/decade) near 1 atmosphere pressure.

e) For gaseous components the concentrations can be related to the pressures through an equation of state. Show that both the Nernst equation and Eq. 3.30 give the same result when the ideal gas law can be used.

f) Would the value of the pressure derivative calculated in Part d be larger at 25°C or at 200°C?

2. Calculate the value of the standard reversible potential for the Au^+/Au^{3+} couple from the values of the Au/Au^+ couple and the Au/Au^{3+} couple.

3. The Weston cell is used as a standard in electrochemical measurements.

$$Cd(Hg) \ / \ CdSO_4 \cdot \frac{8}{3}H_2O \ / \ CdSO_4(sat'd) \ /Hg_2SO_4 \ / \ Hg$$

The overall reaction is

$$Cd(s) \ + \ Hg_2SO_4 \ + \ \frac{8}{3}H_2O(l) \ \rightarrow \ CdSO_4 \cdot \frac{8}{3}H_2O(s) \ + \ 2Hg(l)$$

The reversible potential is

$$E = 1.01845 - 4.05 \times 10^{-5}(T - 20) - 9.5 \times 10^{-7}(T - 20)^2$$

where T is expressed in °C. Calculate the enthalpy of reaction at 25°C.

4. Calculate the equilibrium constant at 25°C for the reaction

$$2Cu^+ \ = \ Cu^{2+} \ + \ Cu$$

5. The lead-acid battery (Planté cell) is used for starting the accessories in all production vehicles. The overall reaction is

$$PbO_2 \ + \ Pb \ + \ 2H_2SO_4 \ = \ 2PbSO_4 \ + \ 2H_2O$$

During discharge the reaction proceeds in one direction, and during the charge cycle, the reaction proceeds in the reverse direction.

a) Write the electrode reactions. Note that sulfuric acid can be written as if it were completely dissociated into hydrogen and sulfate ions.

b) Estimate the standard reversible potential from thermodynamic data. Because free energy data for ions are not always available at the desired concentration, your estimate may be different from tabulated potential data by an amount proportional to a free energy of dilution.

c) Determine the direction of the reaction during discharge.

d) If this cell were operated reversibly at constant temperature and pressure, would it absorb or release heat? Estimate the amount of heat per mole of lead.

6. Another possible reaction on the Pourbaix diagram for the lead-water system is

$$Pb_3O_4 + 2H_2O = 3PbO_2 + 4H^+ + 4e$$

Determine the Nernst equation representation for this reaction, and sketch it on the simplified Pourbaix diagram.

Bibliography

[1] M. Pourbaix, *Atlas of Electrochemical Equilibria in Aqueous Solutions*, 2nd ed., (Houston: NACE, 1974).

[2] S. I. Sandler, *Chemical and Engineering Thermodynamics*, (New York: John Wiley and Sons, 1977), pp. 565-67.

Chapter 4

Phase Equilibrium

In the previous chapter we considered the equilibrium behavior of a cell from a macroscopic point of view. Additional detail can be obtained from an analysis of the thermodynamic behavior of the separate phases. In treating ionic species special consideration must be taken of the electrical state of those species; moreover, a definition of the electrical state of an ion in solution must be developed.

Using the basic framework from non-electrolyte, solution thermodynamics, we will consider the additional complication caused by electrical effects when calculating activity coefficients. Our use of the Nernst equation was based on the simplifying assumption that concentrations could be used in place of activities, and we will be able to assess the magnitude of error caused by this assumption. Further we neglected the effects of bringing two electrolyte phases of different composition into contact. By treating the junction from a thermodynamic perspective, we will be able to estimate the liquid-junction potential arising from the contact between phases.

4.1 Electrochemical Potential

The treatment of electrochemical systems follows the same outline as the treatment of non-electrochemical systems; we will refer to the latter more simply as chemical systems. For a chemical system at constant temperature and pressure, it is the equivalence of the chemical potentials for each species in each phase that constitutes one criterion for equilibrium. This equivalence can be expressed as

$$\mu_i^{\alpha} = \mu_i^{\beta} = \dots \tag{4.1}$$

where μ_i is the chemical potential, which can be defined as

$$\mu_i = \left(\frac{\partial G}{\partial n_i}\right)_{T,P,n_j,j\neq i} \tag{4.2}$$

Note that the symbol n_i refers to the number of moles of a species i; the symbol is used only in partial derivatives to denote the number of moles but it represents the number of electrons in all other contexts. If the criterion of equilibrium stated in Eq. 4.2 is not met, a transfer of material between phases will occur to satisfy that criterion. It is not necessary that each species be present in each phase, but those present must obey Eq. 4.2.

In electrochemical systems the driving force for the transfer of charged species due to differences in electrical potential must be taken into account in a treatment of phase equilibrium. In a chemical system, it is sufficient to have the same chemical environment (at constant temperature and pressure) to maintain equilibrium, i.e., there is no net transfer of material in a system of uniform composition. In electrochemical systems, by contrast, it is not sufficient simply to maintain uniform composition; a uniform potential must also be maintained. In this sense we need to incorporate a measure of the electrical contribution to the chemical potential to properly assess the total potential of a species. It would seem reasonable to simply add together two terms—one for the chemical and another for the electrical contribution—to define an electrochemical potential.

$$\mu_i^{ec} = \mu_i^{chem} + \mu_i^{el} \tag{4.3}$$

For an ionic species in a solution of uniform composition, this type of definition leads to the expected result. If we consider an ionic component i in a solution of uniform composition, but of varying electrical potential, we can calculate the electrochemical potential difference for the ion at different locations within the solution. If we designate two locations α and β, we can use Eq. 4.3 to define the following:

$$\mu_i^{ec,\,\alpha} = \mu_i^{chem,\,\alpha} + \mu_i^{el,\,\alpha} \tag{4.4}$$

$$\mu_i^{ec,\,\beta} = \mu_i^{chem,\,\beta} + \mu_i^{el,\,\beta} \tag{4.5}$$

Because the species is in the same chemical environment in the α and β locations, the chemical contributions must be the same, and the difference in electrochemical potential is calculated by subtracting Eq. 4.4 from Eq. 4.5.

$$\mu_i^{ec,\,\beta} \; - \; \mu_i^{ec,\,\alpha} \; = \; \mu_i^{el,\,\beta} \; - \; \mu_i^{el,\,\alpha} \tag{4.6}$$

All of the difference in electrochemical potential resides in the electrical contribution, as we expect. Any definition of electrochemical potential that did not reduce to this form would be of little use in engineering applications. The contribution to the free energy of the ion is given by the charge number multiplied by the potential; on a molar basis, the contribution is

$$\mu_i^{el} \; = \; z_i F \phi \tag{4.7}$$

and we can express Eq. 4.6 as

$$\mu_i^{el,\,\beta} \; - \; \mu_i^{el,\,\alpha} \; = \; z_i F(\phi^\beta \; - \; \phi^\alpha) \tag{4.8}$$

If we were to transfer a charged species from the α to the β phase, the reversible work required would be given by this equation.

A second case to consider is that of an ionic species in different chemical phases where a potential gradient is also present. Again, a subtraction of Eq. 4.4 from Eq. 4.5 can be performed, but we can no longer eliminate the chemical term because the ions are in different chemical environments.

$$\mu_i^{ec,\,\beta} \; - \; \mu_i^{ec,\,\alpha} \; = \; \mu_i^{chem,\,\beta} \; - \; \mu_i^{chem,\,\alpha} \; + \; \mu_i^{el,\,\beta} \; - \; \mu_i^{el,\,\alpha} \tag{4.9}$$

We have no experimental method of determining the contribution from the chemical environment versus the contribution from the electrical environment for a single ion. We cannot add single ions to a solution—they must be added in neutral combinations; therefore, we cannot measure changes in free energy for incremental additions of one kind of ion.

In practice, we are interested in measuring the electrical state of a solution, which we usually accomplish with a reference electrode. Such an electrode provides a reference ion to which the electrochemical potentials of other species can be related.

The transfer of electrons is possible between certain phases, but at equilibrium the driving force for this transfer must be zero. For aqueous solutions electrons are not present in the solution but can be transferred between electrical—as opposed to ionic—conductors or semiconductors. If we consider electrons moving between two conducting phases α and β, then the equilibrium condition is

$$\mu_e^\beta = \mu_e^\alpha \qquad (4.10)$$

Where reactions between species in phases are possible, the electrochemical potentials of species in the corresponding chemical equation must be equal. For example, consider the reduction of cupric ions from an aqueous phase (β) to a solid phase (α):

$$Cu^{2+} + 2e = Cu \qquad (4.11)$$

The corresponding equilibrium condition is

$$\mu_{Cu^{2+}}^\beta + 2\mu_e^\alpha = \mu_{Cu}^\alpha \qquad (4.12)$$

Symbolically, any reaction can be represented as

$$\sum_i s_i M_i^{z_i} = ne \qquad (4.13)$$

The corresponding equilibrium condition in terms of electrochemical potentials is

$$\sum_i s_i \mu_i = n\mu_e \qquad (4.14)$$

For an electrochemical cell we can relate the electrochemical potentials to the electrical potential of the cell through Eqs. 4.7, 4.10, and 4.13. A cell with an electrolyte of uniform composition is referred to as a cell "without transference," and a cell with concentration gradients is referred to as a cell "with transference." For example, consider the cell

$$\begin{array}{cccc} \alpha & \beta & \delta & \epsilon \end{array}$$

$$Cu \ / \ CuSO_4 \ (c_1) \quad / \quad CuSO_4 \ (c_2) \ / \ Cu \qquad (4.15)$$

In this case, the reversible cell potential is non-zero. Such a cell where differences in concentration alone cause a potential difference is referred to as a concentration cell.

Our convention is to put the more positive electrode on the right. We know from LeChatelier's principle that if c_2 is greater than c_1, there will be a tendency for the cupric ions to deposit on the electrode on the right and to dissolve from the electrode on the left; such a process will tend to equalize the cupric ion concentrations and to drive the cell potential toward zero. More formally, we can look at the Nernst equation for the copper electrode reaction (Eq. 4.11).

$$E = E^\circ - \frac{RT}{nF} \ln\left(\frac{[Cu]}{[Cu^{2+}]}\right)$$

$$E = E^\circ + \frac{RT}{nF} \ln[Cu^{2+}] \tag{4.16}$$

From this equation we see that a higher concentration leads to a higher reversible potential; to be consistent with our convention, c_2 must be greater than c_1.

The same reaction (but in the opposite direction) is occurring at each electrode, and the equilibrium conditions from Eq. 4.14 are

$$\mu^\beta_{Cu^{2+}} + 2\mu^\alpha_e = \mu^\alpha_{Cu} \tag{4.17}$$

$$\mu^\delta_{Cu^{2+}} + 2\mu^\epsilon_e = \mu^\epsilon_{Cu} \tag{4.18}$$

Electrochemical potentials can be evaluated using Eq. 4.8. The reversible potential of the cell is proportional to the difference in electrochemical potential of the electrons in the copper electrodes.

$$FE = z_e F(\phi^\alpha - \phi^\epsilon) = \mu^\alpha_e - \mu^\epsilon_e \tag{4.19}$$

Subtracting Eq. 4.18 from Eq. 4.17 gives us the difference in electrochemical potential of the electrons.

$$\mu^\alpha_e - \mu^\epsilon_e = \frac{\mu^\delta_{Cu^{2+}} - \mu^\beta_{Cu^{2+}}}{2} + \frac{\mu^\alpha_{Cu} - \mu^\epsilon_{Cu}}{2} \tag{4.20}$$

The last term involves the electrochemical potential of copper atoms. For uncharged species the electrical environment does not provide a

driving force for species transport. Because the copper atoms are neutral species in the same chemical environment at each electrode, their
electrochemical potentials are the same in each phase; therefore, the
last term is equal to zero. If the electrodes were chemically different
due to, say, alloying with another metal, the chemical potential of the
copper phases would be different, and the last term could not, in principle, be set equal to zero. Assuming identical electrode composition,
we obtain the following from Eq. 4.20:

$$FE \;=\; \mu_e^{\alpha} \;-\; \mu_e^{\epsilon} \;=\; \frac{\mu_{Cu^{2+}}^{\delta} \;-\; \mu_{Cu^{2+}}^{\beta}}{2} \tag{4.21}$$

To evaluate this expression rigorously, we need to consider the calculation of the electrochemical potentials. For engineering purposes we
would usually obtain an estimate of the cell potential from the Nernst
equation. We know that oxidation occurs at the left electrode, but by
the conventions established in Ch. 2, we treat each electrode reaction
as a reduction.

$$E^{\alpha} \;=\; E^{o} \;-\; \frac{RT}{nF} \; \ln\left(\frac{[Cu]}{[Cu^{2+,\beta}]}\right) \tag{4.22}$$

$$E^{\delta} \;=\; E^{o} \;-\; \frac{RT}{nF} \; \ln\left(\frac{[Cu]}{[Cu^{2+,\delta}]}\right) \tag{4.23}$$

Subtracting the reversible potential of the left-hand electrode from
the potential of the right-hand electrode yields

$$E \;=\; \frac{RT}{nF} \; \ln\left(\frac{[Cu^{2+,\delta}]}{[Cu^{2+,\beta}]}\right) \tag{4.24}$$

By using this equation we are making several assumptions. First, we
have neglected corrections for activity coefficients. Second, because
the concentrations of copper sulfate are different in the two phases,
we have diffusion between the phases. Because diffusion gives rise to
a potential difference and is an irreversible process, we cannot use
the principles of equilibrium thermodynamics to evaluate the potential; moreover, the potential across the two liquid phases depends on
the structure of the junction. Such a potential is referred to as the

liquid-junction potential. The magnitude of the liquid-junction potential is less than 1 mV in cases where the mobilities of the ions in the two phases are all similar and where the concentration differences are small. For large differences in ion mobilities and large concentration differences, the liquid-junction potential is on the order of 10 mV. Because this correction is generally small compared to the reversible cell potential, liquid-junction potentials are often treated as pseudo-equilibrium phenomena. Also, liquid-junction potentials only vary with the type and concentration of ions in the bulk electrolyte; because the bulk electrolyte composition at a liquid junction is often insensitive to changes in cell current, the effect of liquid-junction potentials is frequently included in the reversible cell potential. Evaluation of activity coefficients and liquid-junction potentials is addressed below.

4.2 Activity Coefficients

A standard engineering approximation is to use concentrations in place of activities in thermodynamic calculations. If we are interested in a greater level of accuracy, we need to evaluate the activities of cell components. A conventional approach is to evaluate the activity coefficients and apply these coefficients as correction factors to the concentrations. By relating the activities to the electrochemical potentials, we are in a position to evaluate cell potentials in terms of the phase equilibrium relationships.

To the extent possible we would like to use the familiar framework established for nonelectrolytes, but because of the coulombic effects associated with ions in solution, we cannot use the same models for the computation of activity coefficients. We also need to change our concept of partial molar quantities associated with individual ionic species. Although we can treat single ions theoretically, we cannot measure their properties because we cannot add ions of a single kind to a solution and measure resulting changes. Ions can only be added in neutral combinations such that an electrolyte solution is electrically neutral over any macroscopic volume. Consequently, we need to develop definitions of activities and activity coefficients that can be compared to measured values of these properties.

For nonelectrolytes we frequently perform computations on a mole fraction basis. Although this convention could be maintained in treating electrolyte solutions, it is not customary to do so. The thermodynamic analysis of electrolyte solutions evolved largely from studies of aqueous salt solutions. Because the solubility of most salts in water is limited, the use of mole fractions is less useful than for miscible systems. Although mole fraction, molar, and molal scales have all been employed in electrolyte thermodynamics, the molal scale is convenient for aqueous systems, and we shall use that convention in our treatment of solution thermodynamics. The main disadvantage of the molal scale appears in the treatment of molten salts, where the mass of solvent can approach zero, and the molality approaches infinity. Conversion of activity coefficient values to other scales is, in principle, straightforward.

An activity and an activity coefficient are based on a particular standard state. The activity is the ratio of the fugacity f to that of a standard state fugacity f_i^o:

$$a_i = \frac{f_i}{f_i^o} \tag{4.25}$$

Any convenient state can be used to define the standard state. For example, in non-electrolytes both dilute and concentrated solution standard states are common. These choices correspond to the Henry's law and Lewis and Randall standard states, respectively (Fig. 4.1). For electrolytes formed by the dissolution of salts in liquid solvent, use of the Henry's law to define the standard state is most common. Extrapolation of the Henry's law relation to a concentration of 1 molal defines a hypothetical ideal state to which actual conditions can be referred (Fig. 4.2). In the ideal case the electrolyte dissociates completely into ν ions; therefore, the concentration on the abscissa must be raised to the νth power. For the case shown in the figure, we assume 1:1 electrolyte where $\nu = 2$. The standard state for solids is the material at 25°C. For condensed phases (solids and liquids), there is little effect of pressure changes over a range of tens of atmospheres. For gases the standard state is usually chosen to be the ideal gas state at 1 atm. The correction factor applied to the actual state typically

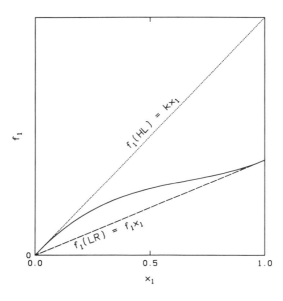

Figure 4.1: Commonly adopted standard states for non-electrolyte solutions. The upper dotted line represents Henry's law and the lower dashed line represents the Lewis and Randall rule.

results in a change in the reversible cell potential on the order of 10 μV at room temperature and pressure. Such a correction is frequently ignored in engineering calculations.

Because we cannot measure individual ion activities, we need to define a measurable activity in terms of individual activities, which we can estimate from theoretical calculations. Values of electrolyte activities can be obtained from osmotic pressure, freezing point depression, and other methods [1]. For a compound that dissociates into ν_+ positive ions and ν_- negative ions according to

$$C_{\nu_+} A_{\nu_-} = \nu_+ C^+ + \nu_- A^- \tag{4.26}$$

we define the mean activity $(a_\pm)$ of the ions in solution as

$$a_\pm^\nu = a_+^{\nu_+} a_-^{\nu_-} \tag{4.27}$$

where $\nu = \nu_+ + \nu_-$. For example, for $ZnCl_2$

$$a_\pm^3 = a_{Zn^{2+}} a_{Cl^-}^2 \tag{4.28}$$

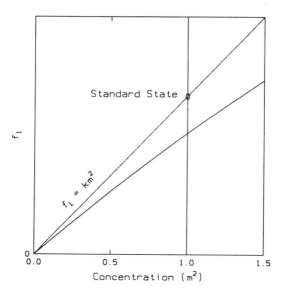

Figure 4.2: Standard state for electrolytes based on behavior in dilute solution.

and the mean activity is

$$a_\pm = \left(a_{Zn^{2+}} a_{Cl^-}^2\right)^{1/3} \tag{4.29}$$

The ionic activity coefficient (γ_i) can be defined as

$$a_i = \gamma_i \, m_i \tag{4.30}$$

Individual ion activity coefficients cannot be measured, and we define the mean activity coefficient as

$$\gamma_\pm^\nu = \gamma_+^{\nu_+} \gamma_-^{\nu_-} \tag{4.31}$$

The mean activity in terms of activity coefficients becomes

$$a_\pm = \left(\gamma_+^{\nu_+} m_+^{\nu_+} \gamma_-^{\nu_-} m_-^{\nu_-}\right)^{1/\nu} \tag{4.32}$$

Because we have chosen the standard state for solutes based on Henry's law, the mean activity coefficient approaches one as the solute concentration approaches zero.

For the special case of a single salt dissolved in a solvent (binary electrolyte), we can use the following relations:

$$m_+ = \nu_+ m \tag{4.33}$$

$$m_- = \nu_- m \tag{4.34}$$

where m_+ and m_- are the molal concentrations of the cation and anion, respectively. The activity then becomes

$$a_\pm = \gamma_\pm \left(\nu_+^{\nu_+} \nu_-^{\nu_-}\right)^{1/\nu} m \tag{4.35}$$

The chemical potential of a non-electrolyte is related to the activity of a species by

$$\mu_i = \mu_i^\circ + RT \ln a_i \tag{4.36}$$

Although we cannot measure the electrochemical potential of an individual ion, we can define such a term. Using the definition of ion activity from Eq. 4.30, we have, from Eq. 4.36,

$$\mu_i = \mu_i^\circ + RT \ln m_i \gamma_i \tag{4.37}$$

The analogous expression for the chemical potential of a binary electrolyte is

$$\mu_{CA} = \mu_{CA}^\circ + RT \ln a_\pm^\nu \tag{4.38}$$

In terms of activity coefficients, Eq. 4.38 becomes

$$\mu_{CA} = \mu_{CA}^\circ + \nu RT \ln(\gamma_\pm m) \tag{4.39}$$

We have not carefully considered the degree of dissociation of a compound in a solvent. For example, Eqs. 4.33 and 4.34 are only valid for complete dissociation of the solute. Even the concept of dissociation in a theoretical sense relies on assumptions regarding the probability of finding an oppositely charged ion in the vicinity of a central ion. Our choice of standard state for a solute is based on an infinitely dilute state, where complete dissociation is assumed. Non-idealities arise from ion association in more concentrated solutions and from other intermolecular and interionic forces. These non-idealities are reflected in the magnitude of the activity coefficient; large percentage deviations from unity indicate that the solution is non-ideal.

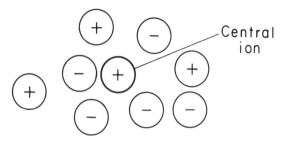

Figure 4.3: Debye-Hückel model of a central, positive ion surrounded by a distribution of ions.

Although the sources of non-idealities are difficult to determine, the overall effect is manifested in calculations of the chemical potential based on measurements of freezing point depression or other colligative properties. In this sense, we do not need to consider the sources of non-idealities to make a thermodynamic calculation. We only need to measure the overall effect and lump all deviations from ideality in the activity coefficient.

4.3   The Debye-Hückel Theory

The first successful, theoretical treatment of interactions among ions was developed by Debye and Hückel in 1923. Their approach was to consider dilute solutions where ions were assumed to be completely dissociated. They ascribed all deviations from ideality to electrical interactions among ions, while ion-solvent interactions were neglected.

They chose a central ion, arbitrarily taken to be positive, surrounded by a distribution of negative and positive ions, as shown in Fig. 4.3. Because the central ion is positively charged, we expect a net excess of negative charges near that ion, but on a macroscopic level electroneutrality is maintained. To make the mathematics more tractable, they replaced the discrete charges (associated with the ions) by a continuum distribution of ions, given by the Boltzmann distribution, as follows:

$$c_i = c_{i,\infty} \exp\left(\frac{-z_i F \phi}{RT}\right) \tag{4.40}$$

where $c_{i,\infty}$ is the bulk concentration of the ith ion.

The potential is treated in the classical electrostatic manner. Because the net excess of ions in the vicinity of the central ion gives rise to a net charge in the medium, Poisson's equation is used to describe the potential distribution

$$\nabla^2 \phi = -\rho/\epsilon \tag{4.41}$$

where ϵ is the permittivity of the medium. Frequently, we refer to the dielectric constant of the medium, which is tabulated in handbooks; it is defined as the permittivity of the medium divided by the permittivity of free space (ϵ/ϵ_0), a dimensionless quantity. The charge density ρ can be expressed in terms of the concentration of the ions as such:

$$\rho = F \sum_i z_i c_i \tag{4.42}$$

Combining Eqs. 4.40, 4.41, and 4.42 yields

$$\nabla^2 \phi = -\frac{F}{\epsilon} \sum_i z_i c_{i,\infty} \exp\left(\frac{-z_i F \phi}{RT}\right) \tag{4.43}$$

Because the model of concentration distribution exhibits spherical symmetry, the Laplacian is cast in spherical coordinates.

$$\frac{1}{r^2}\frac{d}{dr}\left(r^2\frac{d\phi}{dr}\right) = -\frac{F}{\epsilon} \sum_i z_i c_{i,\infty} \exp\left(\frac{-z_i F \phi}{RT}\right) \tag{4.44}$$

This equation is nonlinear, and cannot be solved by elementary techniques; however, we can expand the exponential term in a Maclaurin series by

$$\exp(x) = 1 + x + \frac{x^2}{2} + \dots \tag{4.45}$$

If we assume that the argument of the exponential is small compared to one, then we can closely approximate the exponential with the first two terms. Expanding the exponential in Eq. 4.44 yields

$$\frac{1}{r^2}\frac{d}{dr}\left(r^2\frac{d\phi}{dr}\right) = -\frac{F}{\epsilon}\left(\sum_i z_i c_{i,\infty} - \frac{\sum_i z_i^2 c_{i,\infty} F\phi}{RT}\right) \tag{4.46}$$

Because the solution is electrically neutral overall, the first summation on the right-hand side is equal to zero, and Eq 4.46 becomes

$$\frac{1}{r^2}\frac{d}{dr}\left(r^2\frac{d\phi}{dr}\right) = \frac{\phi F^2 \sum_i z_i^2 c_{i,\infty}}{\epsilon RT} \tag{4.47}$$

This equation is now linear, which is amenable to analytical techniques.

A length, characteristic of the distance over which the field influences the surrounding ions, arises from this analysis. The Debye length is defined as

$$\lambda = \left(\frac{\epsilon RT}{F^2 \sum_i z_i^2 c_{i,\infty}}\right)^{1/2} \tag{4.48}$$

When we incorporate the definition of the Debye length in Eq. 4.47, we obtain

$$\frac{1}{r^2}\frac{d}{dr}\left(r^2\frac{d\phi}{dr}\right) = \frac{\phi}{\lambda^2} \tag{4.49}$$

This ordinary differential equation is second-order and linear. Boundary conditions can be formulated from the following observations: (1) far away from the central ion, the potential decays to zero; and (2) the field at the surface of the central ion is proportional to the charge on the ion. The first boundary condition reflects the fact that the central ion is positively charged, and the potential nearby is positive. From the dilute solution hypothesis we do not expect to encounter separated charges other than those within the influence of the central ion, a few Debye lengths in extent. The second boundary condition is a consequence of Gauss's law: the integral of the outward component of the electric field, multiplied by the permittivity, over the surface of a closed region is equal to the charge enclosed.

$$\oint \epsilon \mathbf{E_f} \cdot ds = \int \rho dV \tag{4.50}$$

At the surface of a single positive charge of radius a, we have a uniform field E_f, and integrating yields

$$\epsilon E_f \left(4\pi a^2\right) = \frac{z_c e}{V} V \tag{4.51}$$

where z_c is the charge on the central ion, e is the unit electric charge (1.6×10^{-19} C), and V is the volume enclosed by the central ion. The magnitude of the electric field is related to the potential gradient by

$$E_f = -\frac{d\phi}{dr} \tag{4.52}$$

Substitution of Eq. 4.52 into Eq. 4.51 gives

$$\frac{d\phi}{dr} = -\frac{z_c e}{4\pi \epsilon a^2} \tag{4.53}$$

Eq. 4.53 is boundary condition 2, and boundary condition 1 is

$$\phi \rightarrow 0 \ \ as \ \ r \rightarrow \infty \tag{4.54}$$

One technique for solving Eq. 4.49 is to make the following substitution:

$$\psi = r\phi \tag{4.55}$$

After substitution and differentiation, Eq. 4.49 becomes

$$\frac{d^2\psi}{dr^2} = \frac{\psi}{\lambda^2} \tag{4.56}$$

The general solution is therefore

$$\psi = A\exp\left(\frac{-r}{\lambda}\right) + B\exp\left(\frac{r}{\lambda}\right) \tag{4.57}$$

where A and B are constants of integration. In terms of the original potential variable ϕ, Eq. 4.57 is

$$\phi = \frac{A}{r}\exp\left(\frac{-r}{\lambda}\right) + \frac{B}{r}\left(\frac{r}{\lambda}\right) \tag{4.58}$$

From the first boundary condition (Eq. 4.54), $B = 0$. Differentiation of Eq. 4.58, evaluated at $r = a$, gives

$$\frac{d\phi}{dr} = -A \left[\frac{\exp(-a/\lambda)}{a^2} - \frac{\exp(-a/\lambda)}{\lambda a} \right] \qquad (4.59)$$

From the second boundary condition (Eq. 4.53), we can evaluate the constant A as

$$A = \frac{z_c e}{4\pi\epsilon} \frac{\exp(a/\lambda)}{1 + a/\lambda} \qquad (4.60)$$

Combining Eqs. 4.58 and 4.60 gives the following potential distribution:

$$\phi = \frac{z_c e}{4\pi\epsilon r} \frac{\exp(a-r)/\lambda}{1 + a/\lambda} \qquad (4.61)$$

Solvated ions are typically a few ångströms in diameter, and the Debye length in dilute aqueous electrolyte is on the order of 10 Å. A plot of the potential distribution from Eq. 4.61 with $a = 3$ Å and $\lambda = 10$ Å appears in Fig. 4.4. The first fraction on the right-hand side represents the potential distribution of a single charge in the medium. The second fraction represents the effect of the excess ions of opposite charge immediately surrounding the central ion. The effect of the ions (predominantly negative) in the immediate vicinity of the central ion is to shield the central ion and lower the potential.

Substitution of Eq. 4.61 into Eq. 4.40 gives an equation for the concentration distribution of the individual species. Negatively charged species are attracted to the central ion and are in excess near it; positively charged ions are repelled, resulting in a deficit near the central ion. In the plot of concentration distribution (Fig. 4.5), the same values were used for a and λ as in the potential distribution calculation. At a distance equal to a few radii from the central ion, the concentration distribution is unaffected by the central ion and the concentrations of positive and negative species are approximately equal.

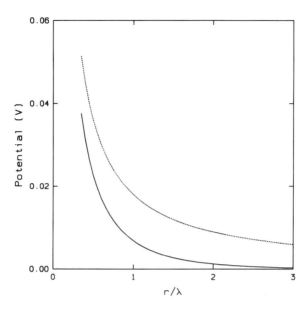

Figure 4.4: Potential distribution as a function of distance from the central ion calculated using Debye-Hückel theory (solid line). For comparison, a bare ion would have the potential distribution indicated by the dashed line.

4.3.1 Evaluation of Activity Coefficients from Debye-Hückel Theory

To estimate a value for the activity coefficient from the calculated potential distribution, we first relate the work required in the potential field starting from an infinitely dilute condition and proceeding reversibly to a more concentrated solution. At constant temperature and pressure, the reversible work is given by the Gibbs free energy. Partial differentiation with respect to the number of moles of an ion gives the value of the electrochemical potential of that ion. In the Debye-Hückel treatment we neglect all but the electrical interactions. Continuing with the same set of assumptions, we attribute all deviations from ideality to coulombic forces; hence, the activity coefficient is directly related to the work required in proceeding from infinite

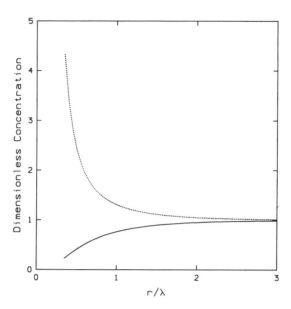

Figure 4.5: Dimensionless concentration distribution $(c_i/c_{i,\infty})$ around a central, positive ion. The upper dotted line represents the concentration of negative ions, and the solid line represents the concentration distribution of the positive ions.

dilution to a finite concentration.

To evaluate the work required to bring ions from infinite dilution to a more concentrated state, we consider an imaginary experiment. We start with ions in a volume of solvent so large that they can be taken as isolated entities with no other nearby ions. Imagine that we can strip the charges from the ions and send the charges far away (where the potential is defined to be zero). The reversible work associated with this process is w_0. The uncharged ions can be transferred to a volume of solvent such that the concentration is the desired finite concentration. because we have assumed that the electrical contributions are dominant, the work associated with the concentration of uncharged species is neglected. In the final step of this imaginary process, we bring the same charges back to the solution and recharge the ions. The work associated with this process is w_1. And, because we have assumed a reversible process, the total work is equal to the change in Gibbs free energy, or

$$G_{el} = w_1 - w_0 \qquad (4.62)$$

The analysis is simplified if we make several assumptions. Eq. 4.61, which expresses potential as a function of distance from the central ion, can be written as

$$\phi = \frac{z_c e}{4\pi \epsilon r} \frac{\exp(a/\lambda) \exp(-r/\lambda)}{1 + a/\lambda} \qquad (4.63)$$

Expanding the exponentials in a Maclaurin series and discarding terms higher than first-order yields

$$\phi = \frac{z_c e}{4\pi \epsilon r} - \frac{z_c e}{4\pi \epsilon \lambda} \qquad (4.64)$$

We also assume that the ions can be fractionally charged, the fraction of charge being equal to ξ at any time. The work to bring a charge from infinity, where the potential is zero, to $r = a$ is given by the charge multiplied by the potential difference.

$$\begin{aligned} w &= q[\phi(r = a) - \phi(r \to \infty)] \\ &= q\,\phi(r = a) \end{aligned} \qquad (4.65)$$

At any time the charge on an ion is

$$dq = z_j e \xi \qquad (4.66)$$

The work required to discharge the infinitely dilute ions is given by the potential at $r = a$, multiplied by the charge. Because there are no surrounding ions in this case, we take only the first fraction in Eq. 4.64, which is

$$w_0 = \int_0^1 \frac{z_j e \xi}{4\pi \epsilon a} z_j e \; d\xi \qquad (4.67)$$

The work required to recharge the ions is not the same because the ions are now in a more concentrated environment, where interaction with surrounding ions is occurring. The potential during the imaginary recharging process is given by Eq. 4.64, evaluated at $r = a$.

Because λ is inversely proportional to the charge, it must be multiplied by the inverse of the fractional charge factor ξ where it appears.

$$w_1 = \int_0^1 \left(\frac{z_j e \xi}{4 \pi \epsilon a} - \frac{z_j e \xi}{4 \pi \epsilon \lambda / \xi} \right) z_j e \, d\xi \qquad (4.68)$$

Thus, the net work is the difference between Eqs. 4.68 and 4.67.

$$
\begin{aligned}
w_1 - w_0 &= \int_0^1 - \left(\frac{z_j e \xi}{4 \pi \epsilon \lambda / \xi} \right) z_j e \, d\xi \\
&= -\frac{z_j^2 e^2}{4 \pi \epsilon \lambda} \int_0^1 \xi^2 d\xi \\
&= -\frac{z_j^2 e^2}{12 \pi \epsilon \lambda} \qquad (4.69)
\end{aligned}
$$

This equation represents the work for a single ion; the total work is the sum of the contributions from each ion. Multiplication of Eq. 4.69 by the number of moles of each species n_j gives the total reversible work, which is the electrical contribution to the Gibbs free energy

$$G_{el} = -\sum_j N n_j \frac{z_j^2 e^2}{12 \pi \epsilon \lambda} \qquad (4.70)$$

where N is Avogadro's number.

As discussed earlier, we cannot measure a single ion activity coefficient, but we can treat it theoretically. Eq. 4.37 is an expression for the electrochemical potential of a single ion in terms of the activity coefficient. For the purposes of using this expression to evaluate the activity coefficient from the Debye-Hückel theory, we can restate Eq. 4.37 as

$$\mu_i = \mu_i^\circ + RT \ln m_i + RT \ln \gamma_i \qquad (4.71)$$

We associate the first two terms on the left-hand side with chemical contributions and assume that all of the non-idealities, arising from electrical effects, are expressed in the activity coefficient.

$$\mu_{i,el} = RT \ln \gamma_i \qquad (4.72)$$

From our expression for the Gibbs free energy (Eq. 4.70), we can evaluate the activity coefficient from Eq. 4.72.

$$\mu_{i,el} = \left(\frac{\partial G}{\partial n_i}\right)_{T,P,n_j} \tag{4.73}$$

$$= -N\frac{z_i^2 e^2}{8\pi\epsilon\lambda} \tag{4.74}$$

In performing the differentiations indicated in Eq. 4.73, we note that λ is inversely proportional to the square root of the concentration (and to n_i). Also, the summation in Eq. 4.70 reduces to the single term in Eq. 4.74 because all of the moles of ions other than the ith kind remain constant. Combining Eqs. 4.72 and 4.74, we obtain our expression for the activity coefficient, which is

$$\ln\gamma_i = -\frac{z_i^2 e^2 N}{8\pi\epsilon\lambda RT} \tag{4.75}$$

Because $F = Ne$, Eq. 4.75 can be written as

$$\ln\gamma_i = -\frac{z_i^2 Fe}{8\pi\epsilon\lambda RT} \tag{4.76}$$

We linearized the exponential terms in Eq. 4.63 to simplify the analysis. Without this simplification an analytical solution can still be obtained, but the manipulations are slightly more involved. If we proceed through the above analysis starting with Eq. 4.64 instead of Eq. 4.77 we obtain

$$\ln\gamma_i = -\frac{z_i^2 Fe}{8\pi\epsilon\lambda RT}\frac{1}{1 + a/\lambda} \tag{4.77}$$

We note that Eqs. 4.76 and 4.77 differ by the factor $1/(1 + a/\lambda)$. The simplified approach does not take into account ion size, and predicts the same value of the activity coefficient for all ions of the same charge in a specified solvent. Equation 4.77 accounts for both ion size a and for concentration changes through the Debye length λ. Because λ is inversely proportional to the square root of concentration, its value becomes larger in more dilute solutions, and a/λ becomes smaller; consequently, Eq. 4.76 is a limiting form of Eq. 4.77 at low concentrations.

Because a number of properties of ionic solutions depends on electrostatic interactions, it is useful to define a function called the ionic strength as

$$I = \frac{1}{2} \sum_i z_i^2 m_i \tag{4.78}$$

The Debye length is also defined in terms of an ionic strength, but it is in molar units. For dilute solutions the molality is proportional to the molarity.

$$m = \frac{c}{\rho_o} \tag{4.79}$$

Eq. 4.77 becomes

$$I = \frac{1}{2\rho_o} \sum_i z_i^2 c_i \tag{4.80}$$

By lumping the constants and using ionic strength, we can clearly see the relationships in the Debye-Hückel expression for the activity coefficient (Eq. 4.77)

$$\ln \gamma_i = -\frac{z_i^2 A\sqrt{I}}{1 + Ba\sqrt{I}} \tag{4.81}$$

where

$$B = \frac{F\sqrt{\rho_o}}{\sqrt{\epsilon RT/2}} \tag{4.82}$$

$$A = \frac{BFe\sqrt{\rho_o}}{8\pi\epsilon/RT}$$

$$= \frac{\sqrt{2}F^2 e\rho_o}{8\pi(\epsilon RT)^{3/2}} \tag{4.83}$$

In extremely dilute solutions the ionic strength is low, and the denominator in Eq. 4.81 approaches a value of one. Under these circumstances, Eq. 4.81 can be expressed in the following limiting form:

$$\ln \gamma_i = -z_i^2 A\sqrt{I} \tag{4.84}$$

The constant A depends only on the type of solvent at a specified temperature, and, according to this equation, the activity coefficient is expected to vary with the square root of ionic strength at low concentrations. Although the dielectric constant of a solution varies with concentration, a value for the pure solvent is used in evaluating the constants A and B in Eqs. 4.82 and 4.84. For aqueous solutions at 25°C, A = 1.18 $(kg/mol)^{1/2}$ and B = 0.329 $(kg/mol)^{1/2}/Å$. If we convert Eq. 4.84 to common logarithms and incorporate the conversion factor in the constant, Eq. 4.84 becomes

$$\log \gamma_i = -z_i^2 A' \sqrt{I} \qquad (4.85)$$

where $A' = 0.51$.

Because all of the constants are known in Eq. 4.85, we can test it against experimental data at low concentrations. From Eq. 4.31, the mean activity coefficient can be expressed as

$$\log \gamma_{\pm} = \frac{\nu_+ \log \gamma_+ + \nu_- \log \gamma_-}{\nu_+ + \nu_-} \qquad (4.86)$$

If we substitute expressions for γ_+ and γ_- from Eq. 4.85 into Eq. 4.86 and recognize that $|z_+\nu_+| = |z_-\nu_-|$, we obtain

$$\log \gamma_{\pm} = -A' z_+ z_- \sqrt{I} \qquad (4.87)$$

Calculated values of the activity coefficient from this equation are in agreement with experimental data. The agreement is surprisingly good in light of the assumptions that were made in the derivation (see Prob. 4). In fact, all of the predictions of the model, embodied in Eq. 4.87, are confirmed at low electrolyte concentrations. A plot of the logarithm of the activity coefficient vs. $I^{1/2}$ at low concentrations demonstrates that the plot: (1) is linear; (2) the slope is usually within 1% of the predicted value; and (3) ions of the same charge number fall on the same line. Furthermore, because the constant A is inversely proportional to the dielectric constant of the medium, the slope of the plot should increase with decreasing dielectric constant. This prediction is also confirmed from measurements in other solvents.

A plot of activity coefficients over a range of concentrations (Fig. 4.6) reveals that the monotonic decrease in activity coefficient values

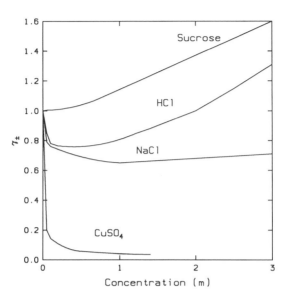

Figure 4.6: Measured values of activity coefficients for solutions of electrolytes at 25°C.

is not realized at higher concentrations for many electrolytes. Activity coefficients derived from the Debye-Hückel theory generally agree with experimental values within about 10% for electrolyte concentrations less than 0.1 M. A comparison between predicted and experimental values appears in Table 4.1. Use of the more accurate form of the model (Eq. 4.81) allows an extension of the range of validity, typically by a factor of about 5. When using this equation, the value of the ion-size parameter is about the right order of magnitude, but it is generally treated as a fitting parameter.

A minimum effective ion size should be the distance of closest approach of the bare ions, which should be about equal to the sum of the crystal radii of the ions. A maximum size should be equal to the sum of the radii of the hydrated ions. In practice, one can measure an activity coefficient at one concentration and calculate a value of the ion-size parameter for use at other concentrations. Values of the ion-size parameter are typically 3 to 5 Å.

In general, the activity coefficient decreases with increasing con-

Concentration m	$\gamma_\pm$ Calculated	$\gamma_\pm$ Experimental
0.001	0.965	0.965
0.005	0.927	0.927
0.01	0.901	0.902
0.05	0.816	0.821
0.1	0.769	0.778
1.0	0.353	0.657

Table 4.1: Comparison of observed values with activity coefficients calculated from Debye-Hückel theory for aqueous sodium chloride solutions at 25°C. An ion-size parameter of 4 Å was used in the calculations.

centration at low concentrations, as shown in Fig. 4.6. The value of the activity coefficient usually passes through a minimum near 1 M concentration. The decrease in the magnitude of the activity coefficient with concentration is predicted by the Debye-Hückel theory. Long-range coulombic forces reduce the electrochemical potential, and the activity coefficient, of ions. At higher concentrations an increase in the activity coefficient is not predicted by the theory.

Several qualitative explanations of activity coefficient behavior at higher concentrations can be considered. We have neglected a number of factors in the formulation of the Debye-Hückel model, including ion-solvent interactions, solvation of ions, association of ions, and variation of dielectric constant with concentration. For non-electrolytes, interactions with solvent tend to increase the activity coefficient with increasing concentration, as illustrated for sucrose in Fig. 4.6. Solvation effects can also be expected to increase the activity coefficient by binding the solvent. Only the quantity of unbound solvent acts as a diluent. The effective concentration is, therefore, much higher, which causes an increase in the activity coefficient.

4.3.2 Modifications to the Debye-Hückel Theory

The Debye-Hückel model is obviously too simple for use in calculating activity coefficients over a concentration range of general interest for engineering purposes. Its real value is that it gives us insight into the thermodynamic behavior of dilute ionic solutions. Also, it serves as a basis for more complicated models and correlations. Although a number of theoretical refinements have subsequently been carried out, they have only been successful in suggesting the proper forms for quantitative evaluation of activity coefficients.

A common approach is to add terms to the Debye-Hückel expression. For example, the following relationship has been proposed [2]:

$$- \log \gamma_\pm \; = \; \frac{A_m I^{1/2}}{1 + B_m I^{1/2}} \; + \; 2B_2 I \; + \; \frac{3}{2} \, C I^2 \; + \; \frac{4}{3} D I^3 \qquad (4.88)$$

One of the more successful single parameter expressions was developed by Bromley [3]

$$\log \gamma_\pm \; = \; \frac{-A_m |z_+ z_-| I^{1/2}}{1 + a_i I^{1/2}} \; + \; \frac{(0.06 + 0.6 B_m)|z_+ z_-| I}{[1 + (1.5 I/|z_+ z_-|)]^2} \; + \; B_m I \quad (4.89)$$

where a_i is a function of the type of electrolyte and B_m depends on the temperature and type of electrolyte. When correlated with data at an ionic strength of 6, this expression generally yields an activity coefficient within $\pm \, 5\%$ of the experimental value. A review of available correlations has been presented by Horvath [4]. Additional data on multicomponent systems can be found in the handbook by Zemaitis et al. [5]. Techniques for estimating activity coefficients at different temperatures, and correlations for multicomponent mixtures of electrolytes are referred to in that review. A table of activity coefficients appears in Appendix D.

For solutions more dilute than 0.1 M, the values of activity coefficients are approximately equal on any of the common scales: molar, molal, and mole fraction. At higher concentrations the activity coefficients are simply related. For binary electrolytes these relations are

$$\gamma_{m,\pm} \; = \; \gamma_{c,\pm} \, \frac{c}{m \rho_0} \qquad\qquad (4.90)$$

$$= \; \frac{\gamma_{x,\pm}}{1 + 0.001 \nu M_0 m} \qquad (4.91)$$

where M is the molecular weight, the subscript 0 refers to pure solvent, and ν represents the number of ions derived from a molecule of salt.

We can apply activity coefficient corrections in a straightforward manner in thermodynamic calculations. For example, we can apply these corrections to the Daniell cell, shown below, when the electrolyte concentrations are each one molal.

$$\text{Zn / ZnSO}_4 \text{ / CuSO}_4 \text{ / Cu} \tag{4.92}$$

The Nernst equation is

$$E \;=\; E^\circ \;-\; \frac{2.303\ RT}{2F}\ \log\left(\frac{a^2_{\text{ZnSO}_4,\pm}}{a^2_{\text{CuSO}_4,\pm}}\right) \tag{4.93}$$

At 1 molal concentration at 25°C the activity coefficients are: $\gamma_\pm = 0.047$ for $CuSO_4$ and $\gamma_\pm = 0.045$ for $ZnSO_4$. From Eq. 4.35 the mean activities can be related to the activity coefficients. For both $ZnSO_4$ and $CuSO_4$ $\nu_+ = \nu_- = 1$. The mean activity for each of the electrolytes at 1 molal concentration is numerically equal to the value of the activity coefficient.

$$a_\pm \;=\; \gamma_\pm \tag{4.94}$$

With the substitution of Eq. 4.94 into Eq. 4.93, and the standard reversible potential calculated previously, we have

$$E \;=\; 1.100 \;-\; 0.059 \log\left(\frac{0.045}{0.047}\right) \tag{4.95}$$
$$ \;=\; 1.101 \text{ V} \tag{4.96}$$

In this example we accounted for activity coefficient corrections but neglected liquid-junction potentials that arise from the contact of solutions containing ions of differing mobility. If we are interested in obtaining a better estimate of cell potential than is available from the Nernst equation, we should attempt to estimate both activity coefficient corrections and liquid-junction potentials simultaneously. As previously mentioned, the liquid-junction potential is not a thermodynamic phenomenon, but it is often conveniently included with other thermodynamic components of cell potential.

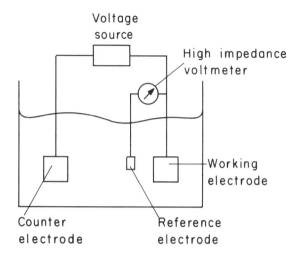

Figure 4.7: Measurement of a working electrode potential with respect to a reference electrode.

4.4   Potential in Solution

In the development of the Debye-Hückel theory, we were able to use an electrostatic basis for the definition of potential in solution. In the analysis of cells with electrolyte of uniform composition, we saw that the chemical contributions to the electrochemical potential cancelled, and we were left with differences in electrical potential (Eq. 4.8). If we were only to treat cells of uniform composition, we would be able to continue using our electrostatic basis for electrical potential; it is in the treatment of cells with electrolytes of varying compositions that we need to be more careful in defining the potential.

For most engineering purposes, the electrical state of an electrode of interest (the working electrode) is measured with a reference electrode (Fig. 4.7). Because the reactions occurring at the interface of the reference electrode and the electrolyte are known, an approximate analysis of the electrochemical cell formed by the reference-working electrode is usually straightforward. An analysis using the Nernst equation is often satisfactory to obtain a relative value of an electrode potential within 10 mV or so.

In considering the level of accuracy required in an analysis, we should first consider the application. Knowledge of the potential at the working electrode might be used for several purposes: (1) to determine electrical losses; (2) to determine a local current density from known current-potential behavior; (3) to maintain a protection potential on a metal in anodic or cathodic protection; and (4) to estimate kinetic parameters.

If we make potential measurements for any of these applications, we are determining potentials under non-equilibrium conditions. Our calculations might be useful in subtracting out the equilibrium components, which might also be determined using a reference electrode. In common industrial systems, changes in electrode surfaces—due to catalyst poisoning, deposit formation, or other causes—can result in changes on the order of 10 mV. In making a measurement with a reference electrode, several other operational factors contribute to the error. Because the reference electrode cannot be placed precisely on the surface of the working electrode, there is a potential difference owing to the resistance of the solution between the reference and working electrodes. This difference is commonly referred to as ohmic drop and measurement or calculation to allow for this difference is referred to as ohmic compensation.

In simple, symmetric electrochemical cells, the ohmic drop can be computed from an analytical solution of the potential distribution in the electrolyte. These techniques are reviewed in Ch. 7. For example, the potential variation in the electrolyte between two large, plane parallel electrodes is a simple linear function.

$$\Delta\phi_{ohm} = i\, d/\kappa \qquad (4.97)$$

where d is the distance between the reference and working electrodes. In this geometry we can calculate the ohmic drop to a reference electrode 5 mm from the working electrode. If the current density is 10 mA/cm^2 and the solution conductivity is 0.1 ohm^{-1}-cm^{-1}, then the ohmic drop is 50 mV. Ohmic drop varies with local current density, electrolyte conductivity, and reference electrode placement. Both experimental and computational procedures have been used in correcting for ohmic drop [6]. Factors affecting the choice of reference electrode locations are discussed in Ch. 5.

It is only when we need a higher level of accuracy—usually for research purposes—that we need to take into account phenomena that we have neglected so far. In detailed investigations we would take precautions that would allow us to obtain higher accuracy. For the study of solid electrodes in liquid electrolyte such precautions might include purifying the electrolyte, maintaining strict temperature control, using pure samples, polishing the electrode, and deareating the electrolyte. Under these conditions we might be able to make measurements of electrode potential within a few millivolts or less of the true value. A more detailed analysis might then be warranted.

Several different definitions of potential are used in detailed analyses of electrochemical cells. When we combine Eqs. 4.3 and 4.7 we obtain

$$\mu_i^{ec} = \mu_i^{chem} + z_i F \phi \qquad (4.98)$$

Because we cannot uniquely apportion the components of the electrochemical potential, we need to adopt an arbitrary reference potential for one ion to which all others will be referred. Although a number of different possibilities exist, the following definition has proved generally useful:

$$\mu_n = RT \ln m_n + z_n F \phi \qquad (4.99)$$

where the species n is the reference ion, which is often taken to be a non-reacting counterion. This definition is referred to as the quasi-electrostatic potential [7]. It has the advantages that it can be directly related to the electrochemical potential, and it reduces to the electrostatic form for solutions of uniform composition. In any application we are only interested in neutral combinations of ions; therefore, when taking a difference in electrochemical potentials between species i and n, we need to multiply the electrochemical potential by the factor z_i/z_n to account for any difference in charge on the ions. From Eq. 4.37 we have

$$\mu_i - \frac{z_i}{z_n}\mu_n = \mu_i^\circ - \frac{z_i}{z_n}\mu_n^\circ + RT[\ln(\gamma_i m_i) - \frac{z_i}{z_n}\ln(\gamma_n m_n)] \qquad (4.100)$$

Substituting Eq. 4.99 into Eq. 4.100 yields

$$\mu_i = \mu_i^\circ - \frac{z_i}{z_n}\mu_n^\circ + RT \ln m_i + z_i F \phi + RT(\ln \gamma_i - \frac{z_i}{z_n}\ln \gamma_n) \qquad (4.101)$$

After we make a detailed analysis of a cell in terms of the phase equilibria, we should be able to use the definition of the potential embodied in Eq. 4.101 to derive a more accurate expression for the cell potential than is given by the Nernst equation.

We can treat the phase equilibrium for the Daniell cell.

$$\alpha \qquad \beta \qquad \delta \qquad \epsilon$$

$$\text{Zn} \;/\; \text{ZnSO}_4 \;(m=1) \;\;/\;\; \text{CuSO}_4 \;(m=1) \;/\; \text{Cu} \qquad (4.102)$$

The reactions are

$$\text{Zn}^{2+} + 2e = \text{Zn} \qquad (4.103)$$
$$\text{Cu}^{2+} + 2e = \text{Cu} \qquad (4.104)$$

Corresponding phase equilibrium relationships from Eq. 4.14 are

$$\mu^{\beta}_{\text{Zn}^{2+}} + 2\mu^{\alpha}_{e} = \mu^{\alpha}_{\text{Zn}} \qquad (4.105)$$
$$\mu^{\delta}_{\text{Cu}^{2+}} + 2\mu^{\epsilon}_{e} = \mu^{\epsilon}_{\text{Cu}} \qquad (4.106)$$

Applying Eq. 4.8 to this system yields

$$FE = z_e F(\phi^{\alpha} - \phi^{\epsilon}) = \mu^{\alpha}_{e} - \mu^{\epsilon}_{e} \qquad (4.107)$$

The difference in electrochemical potential of the electrons can be determined from Eqs. 4.105 and 4.106.

$$\mu^{\alpha}_{e} - \mu^{\epsilon}_{e} = \frac{1}{2}(\mu^{\alpha}_{\text{Zn}} - \mu^{\beta}_{\text{Zn}^{2+}} + \mu^{\delta}_{\text{Cu}^{2+}} - \mu^{\epsilon}_{\text{Cu}}) \qquad (4.108)$$

For pure phases (Zn and Cu) we use the standard chemical potentials, and for the ionic species we use the expression for the electrochemical potential in terms of the electric potential (Eq. 4.101). We choose the sulfate ion as our reference species, and Eq. 4.108 becomes

$$
\begin{aligned}
FE = \;& \frac{1}{2}(\mu^{o}_{\text{Zn}} - \mu^{o}_{\text{Cu}}) + \frac{1}{2}\{\mu^{o}_{\text{Cu}^{2+}} - \frac{2}{-2}\mu^{o}_{\text{SO}_4^{2-}} + z_{\text{Cu}^{2+}} F\phi^{\delta} \\
& + RT[\ln\gamma^{\delta}_{\text{Cu}^{2+}} - \frac{2}{-2}\ln\gamma^{\delta}_{\text{SO}_4^{2-}}] + RT\ln m^{\delta}_{\text{Cu}^{2+}}\} \\
& - \frac{1}{2}\{\mu^{o}_{\text{Zn}^{2+}} - \frac{2}{-2}\mu^{o}_{\text{SO}_4^{2-}} + z_{\text{Zn}^{2+}} F\phi^{\beta} \\
& + RT[\ln\gamma^{\beta}_{\text{Zn}^{2+}} - \frac{2}{-2}\ln\gamma^{\beta}_{\text{SO}_4^{2-}}] + RT\ln m^{\beta}_{\text{Zn}^{2+}}\} \qquad (4.109)
\end{aligned}
$$

All of the terms with the superscript o are concentration-independent and can be combined into the standard electrode potential E^o.

$$FE = FE^o + \frac{1}{2}RT\ln\frac{\gamma^\delta_{Cu^{2+}}\gamma^\delta_{SO_4^{2-}}m^\delta_{Cu^{2+}}}{\gamma^\beta_{Zn^{2+}}\gamma^\beta_{SO_4^{2-}}m^\beta_{Zn^{2+}}} + F(\phi^\delta - \phi^\beta) \qquad (4.110)$$

The last term is the potential difference between the two electrolyte phases, which is the liquid-junction potential. If we neglect activity coefficient corrections and the liquid-junction potential, we are left with the Nernst equation. Usually, concentrations are expressed in mol/L, and a conversion to that unit would make a significant difference in concentrated solutions.

4.5   Liquid-junction Potential

We have seen from the previous analysis that we have a potential difference between different electrolyte phases in an electrochemical cell. In a strictly thermodynamic analysis of an electrochemical cell, we are not concerned with the detailed construction of the cell. The geometry of the electrodes, quantity of electrolyte, and interelectrode gap are not relevant to the analysis. Because the liquid-junction potential is a diffusion potential, rather than a thermodynamic potential, we need to account for any factors that affect the rates of diffusion across the interface of the two electrolyte phases. The exact details of the interfacial structure vary from cell to cell, but a number of models have been developed to account for a range of behavior.

Although the liquid-junction potential is a non-thermodynamic quantity, its inclusion in the measurement of cell potential—even when a cell is considered to be at rest—cannot be avoided; therefore, it is usually included with other thermodynamic quantities. Other sources of irreversible potential loss, such as ohmic drop and overpotential, can be minimized by reducing current density to a sufficiently low level; consequently, these current-dependent potential losses are calculated separately.

To gain insight into the origin of a liquid-junction potential, we consider the interface between two different electrolytes, e.g., HCl and

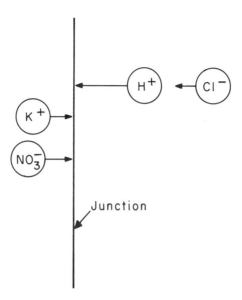

Figure 4.8: Interface between two different electrolyte solutions. The size of the arrows indicates the relative diffusivity. Because of the greater diffusivity of hydrogen ions, the orientation of the hydrogen ions gives rise to a liquid-junction potential.

KNO_3 (Fig. 4.8). Because of the concentration gradients between the two phases, there is a tendency for the HCl to diffuse to the left and for the KNO_3 to diffuse to the right. If these were non-electrolytes, the species would tend to diffuse as ordinary molecules, and the diffusion rate could be estimated from the basic laws of diffusion. In the case of ionic species, the molecules dissociate, and we need to consider the diffusion of each ion separately. Because the electrical forces between ions of different signs are so large, the possibility of any significant deviation from electroneutrality on a macroscopic basis is precluded; however, preferential orientations of ions on a microscopic basis is possible.

Because of the deficiency of both hydrogen and chloride ions on the left-hand side, the HCl tends to diffuse to the left. The rate of diffusion for an ionic species is proportional to its diffusivity. For HCl the hydrogen ions are more mobile than the chloride ions. In dilute aqueous electrolyte near room temperature, the diffusivity of a hydrogen ion is approximately 9×10^{-5}, but that of a chloride ion is only 2×10^{-5}. If these species were uncharged, the hydrogen ions would diffuse more rapidly, but because the species are charged, separation is constrained by electroneutrality. We might think of the hydrogen ions as moving ahead of the chloride ions, and consequently maintaining a preferential orientation. Potassium and nitrate ions have about the same diffusivity, and they should diffuse at the same rate. It is the difference in diffusivity and resulting preferential orientation of two different ions that give rise to a potential difference.

We can also consider a second interface where we have KCl on one side and KNO_3 on the other. Again we expect to have diffusion across the interface, but in this case the diffusivities of all species are about the same as that of the chloride ion; consequently, we expect less of a tendency for the ionic species to be oriented near the interface, and a lower value of a liquid-junction potential. This example is in agreement with experimental observations.

As we might also expect intuitively, concentrations of the species on each side of the interface influence the magnitude of the liquid-junction potential. For example, if we consider an HCl/KCl interface, we would expect to see a larger value of the liquid-junction potential if the concentration ratios of HCl:KCl were 1:1 rather than 1:10, i.e., the liquid-junction potential is strongly influenced by the species in higher concentration. This phenomenon forms the basis of a technique for reducing the magnitude of the liquid-junction potential. If we use a salt such as KCl, where the diffusivities of the cation and anion are approximately equal, the interface formed from a concentrated solution of KCl and another salt reduces the liquid-junction potential. This reduction implies that if we place an intervening phase of concentrated KCl in a cell having a high liquid-junction potential, we would expect the sum of the liquid-junction potentials to be reduced.

We can develop a quantitative model of the potential difference across an electrolyte interface from a transport or thermodynamic point of view. Because we have developed a thermodynamic framework, we will proceed from that perspective. First, we consider a simple cell with an interface formed by the same electrolyte at different concentrations. Afterward, we will generalize the concepts to include different electrolytes and develop a series of expressions that are useful in making calculations at different levels of accuracy.

4.5.1 Single Component Solutions

We first consider an interface between aqueous solutions of KCl at two different concentrations. If we retain the sign conventions established in Ch. 2, positive current always flows from left to right.

$$\alpha \qquad\qquad \beta$$

$$KCl\ (c_1) \quad / \quad KCl\ (c_2) \qquad\qquad (4.111)$$

The passage of one Faraday of charge through a cell containing this interface results in the movement of t_+ cations (potassium ions in this case) from left to right and t_- anions to the left. If transference numbers are assumed to be independent of concentration, then we have

$$\Delta G\ =\ -FE_l \qquad\qquad (4.112)$$

where E_l is the liquid-junction potential. The total change in free energy is proportional to the difference in electrochemical potential of each ionic species, or

$$\Delta G\ =\ t_+(\ \mu_+^\beta - \mu_+^\alpha\ )\ +\ t_-(\mu_-^\alpha\ -\ \mu_-^\beta) \qquad\qquad (4.113)$$

Electrochemical potentials can be expressed in terms of activities according to Eq. 4.36.

$$\Delta G\ =\ t_+ RT \ln \frac{a_+^\beta}{a_+^\alpha}\ +\ t_- RT \ln \frac{a_-^\alpha}{a_-^\beta} \qquad\qquad (4.114)$$

The combination of Eqs. 4.112 and 4.114 gives an expression for the liquid-junction potential in terms of ion activities.

$$E_l\ =\ -t_+ \frac{RT}{F} \ln \frac{a_+^\beta}{a_+^\alpha}\ +\ t_- \frac{RT}{F} \ln \frac{a_-^\beta}{a_-^\alpha} \qquad\qquad (4.115)$$

The sum of the transference numbers is equal to one; therefore, this equation can also be expressed as

$$E_l = -2t_+ \frac{RT}{F} \ln \frac{a_-^\beta}{a_-^\alpha} + \frac{RT}{F} \ln \frac{a_-^\beta}{a_-^\alpha} \qquad (4.116)$$

where the a's refer to the mean activity of the KCl. If we assume that the ratio of the activities of the anions is approximately equal to the mean activities of the electrolyte, then Eq. 4.116 becomes

$$E_l = (1 - 2t_+) \frac{RT}{F} \ln \frac{a_-^\beta}{a_-^\alpha} \qquad (4.117)$$

$$= (t_- - t_+) \frac{RT}{F} \ln \frac{a_-^\beta}{a_-^\alpha} \qquad (4.118)$$

This expression was derived for 1:1 electrolyte and is only valid for solutions of that type. Eq. 4.118 illustrates that the liquid-junction potential is proportional to the difference in transference numbers between the anion and cation; the transference number, in turn, is proportional to ion mobility. The sign dependence is also what we intuitively expect. For example, if we have a higher concentration in the α phase of an electrolyte having a more mobile cation, then the pre-logarithmic term and the logarithmic term in Eq. 4.118 are both negative, and the liquid-junction potential makes a positive contribution to the cell potential.

4.5.2 Multicomponent Solutions

More generally, we can have a number of different ions at different concentrations, which form a junction. The increase in Gibbs free energy during the passage of one Faraday for the transfer of an ionic species from a concentration c_i to a concentration $c_i + dc_i$ is

$$dG = \frac{t_i}{z_i} [(\mu_i + d\mu_i) - \mu_i]$$

$$= \frac{t_i}{z_i} d\mu_i \qquad (4.119)$$

The total change in free energy for the passage of one Faraday is the sum of the changes for all species.

$$\Delta G = \sum_i \frac{t_i}{z_i} d\mu_i$$

$$= \sum_i \frac{t_i}{z_i} d\ln a_i \tag{4.120}$$

To approximate the transfer of ions as a reversible process with small differences in concentration in the junction region, we model the interface as a series of layers of electrolyte. Each layer contains electrolyte that differs in concentration only infinitesimally from the adjacent layer. For the transfer of ions between adjacent layers, the differential liquid-junction potential can be determined from Eqs. 4.112 and 4.120.

$$dE_l = -\frac{RT}{F} \sum_i \frac{t_i}{z_i} d\ln a_i \tag{4.121}$$

The integration of Eq. 4.121 gives

$$E_l = -\frac{RT}{F} \int_\alpha^\beta \sum_i \frac{t_i}{z_i} d\ln a_i \tag{4.122}$$

where the limits of integration represent the activities for each species in the two phases. This general form of the equation calculates the liquid-junction potential. There are a number of special cases for which integration reduces this equation to a form suitable for engineering calculations.

If the electrolyte on both sides of a junction is the same binary electrolyte at different concentrations, then Eq. 4.122 becomes

$$E_l = -\frac{RT}{F} \int_\alpha^\beta \frac{t_+}{z_+} d\ln a_+ + \frac{RT}{F} \int_\alpha^\beta \frac{t_-}{z_-} d\ln a_- \tag{4.123}$$

If we assume that the transference numbers are independent of concentration, then we obtain the following integrated equation:

$$E_l = -\frac{t_+}{z_+} \frac{RT}{F} \ln \frac{a_+^\beta}{a_+^\alpha} + \frac{t_-}{z_-} \frac{RT}{F} \ln \frac{a_-^\beta}{a_-^\alpha} \tag{4.124}$$

For a 1:1 electrolyte this expression reduces to Eq. 4.115.

4.5.3   The Henderson Equation

Eq. 4.122 can be integrated by elementary means for multicomponent systems if we make the following additional assumptions: (1) the activities can be replaced by concentrations; and (2) the concentration gradients in the junction region are linear. The transference number was defined previously as

$$t_i \;=\; \frac{z_i^2 u_i c_i}{\sum_j z_j^2 u_j c_j} \tag{4.125}$$

If we express $d\ln c_i$ as dc_i/c_i and substitute Eq. 4.125 into 4.122, we have

$$E_l \;=\; -\frac{RT}{F} \int_\alpha^\beta \frac{\sum_i z_i u_i \; dc_i}{\sum_j z_j^2 u_j c_j} \tag{4.126}$$

Linear gradients for each species can be expressed as

$$c_i \;=\; c_i^\alpha \;+\; (c_i^\beta \;-\; c_i^\alpha)x \tag{4.127}$$

where x is the fractional distance across the junction with the origin in the α phase, i.e., $x = 0$ just to the left of the junction region and $x = 1$ just to the right of the junction region. The differential concentration is then proportional to the differential fractional distance.

$$dc_i \;=\; (c_i^\beta \;-\; c_i^\alpha)dx \tag{4.128}$$

Substituting Eqs. 4.127 and 4.128 into Eq. 4.126 yields

$$E_l \;=\; -\frac{RT}{F} \int_\alpha^\beta \frac{\sum_i z_i u_i (c_i^\beta \;-\; c_i^\alpha)dx}{\sum_i z_i^2 u_i [c_i^\alpha \;+\; (c_i^\beta \;-\; c_i^\alpha)x]} \tag{4.129}$$

In this equation we have arbitrarily changed all index subscripts to i because we are summing over all species. To express this summation in more compact form we let

$$H \;=\; \sum_i z_i u_i (c_i^\beta \;-\; c_i^\alpha) \tag{4.130}$$

$$J^\alpha = \sum_i z_i^2 u_i c_i^\alpha \tag{4.131}$$

$$J^\beta = \sum_i z_i^2 u_i c_i^\beta \tag{4.132}$$

Substituting Eqs. 4.130 through 4.132 into Eq. 4.129 gives

$$E_l = -\frac{RT}{F} \int_\alpha^\beta \frac{H\,dx}{J^\alpha + (J^\beta - J^\alpha)x} \tag{4.133}$$

Carrying out the integrations, we obtain

$$E_l = -\frac{RT}{F} H \frac{\ln(J^\beta/J^\alpha)}{J^\beta - J^\alpha} \tag{4.134}$$

This expression is the Henderson equation for the liquid-junction potential. It is valid for dilute solutions, where activity coefficient corrections can be ignored. The expression is further restricted to junctions where linear gradients form a so-called continuous-mixture junction.

This expression can be simplified for the case of a single, symmetric electrolyte ($z = z_- = z_+$) at different concentrations.

$$E_l = -\frac{RT}{|z|F} \frac{(u_+ - u_-)}{(u_+ + u_-)} \ln \frac{c^\beta}{c^\alpha} \tag{4.135}$$

For 1:1 electrolyte the transference number of the cation can be expressed as $u_+/(u_+ + u_-)$, and this equation reduces to

$$E_l = (2t_+ - 1)\frac{RT}{F} \ln \frac{c^\alpha}{c^\beta} \tag{4.136}$$

This equation corresponds to Eq. 4.117 with concentrations in place of activities.

The Henderson equation can be simplified in the special case where two 1:1 electrolytes with a common ion are at the same concentration. If we consider an electrolyte with a common anion, Eq. 4.134 becomes

$$\begin{aligned}
E_l &= \frac{RT}{F} \ln \frac{u_{+,\alpha} + u_-}{u_{+,\beta} + u_-} \\
&= \frac{RT}{F} \ln \frac{\Lambda_\alpha}{\Lambda_\beta}
\end{aligned} \tag{4.137}$$

Junction	E_l (mV) Calculated	E_l (mV) Experimental
HCl/KCl	28.52	26.78
HCl/LiCl	36.14	34.86
KCl/LiCl	7.62	8.79
KCl/NaCl	4.86	6.42
NaCl/NH$_4$Cl	-4.81	-4.21

Table 4.2: Comparison of calculated and observed values for liquid-junction potentials. All solutions are 0.1 M.

where the Λ's are the equivalent conductances of the two solutions. This expression is the Lewis and Sargent formula. A comparison between calculated and experimental results [8] appears in Table 4.2. Liquid-junction potentials are typically a few tens of millivolts in dilute electrolyte when transference numbers of ions differ significantly. Further refinement in the calculation of liquid-junction potentials has been carried out. A number of different models for junctions have been proposed including those for flowing electrolyte [7]. Including activity coefficient corrections and concentration dependence of transference numbers in a numerical integration of Eq. 4.122 should give improved accuracy.

Reduction in the magnitude of the liquid-junction potential is often accomplished by connecting two different electrolyte solutions with a salt bridge. Usually, the bridge contains a concentrated solution of an equitransferent electrolyte, i.e., one having similar cation and anion transference numbers such as KCl. Because the concentration of each species enters the Henderson equation, the net effect of a higher concentration of an equitransferent ion is to reduce the liquid-junction potential of the cell. This effect is illustrated in Fig. 4.9 for the HCl/KCl junction [9]. Initially, without a salt bridge, the liquid-junction potential is approximately 29 mV. When a bridge containing concentrated KCl is included in the cell, the liquid-junction potential is reduced to a value below 3 mV.

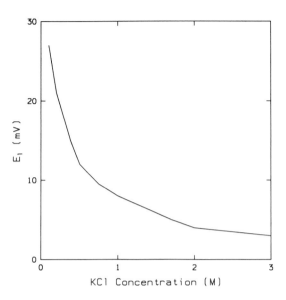

Figure 4.9: The effect of concentration in a KCl salt bridge on the liquid-junction potential.

4.6 Problems

1. The cell for a lead-acid battery can be represented by

$$\begin{array}{cccccc} \alpha & \beta & \delta & \epsilon & \zeta & \eta \end{array}$$
$$\text{Pb}/ \quad \text{PbSO}_4/ \quad \text{H}_2\text{SO}_4, \text{H}_2\text{O}/ \quad \text{PbSO}_4/ \quad \text{PbO}_2/ \quad \text{Pb}$$

The electrode reactions are

$$\text{PbO}_2 + \text{SO}_4^{2-} + 4\text{H}^+ + 2e = \text{PbSO}_4 + 2\text{H}_2\text{O}$$
$$\text{PbSO}_4 + 2e = \text{Pb} + \text{SO}_4^{2-}$$

a) Write the equilibrium relationships in terms of electrochemical potentials.

b) Derive an expression for the overall cell potential in terms of electrochemical potentials.

c) Derive the Nernst equation from the equilibrium relationships.

d) Calculate the standard cell potential.

2. The following cell has been considered for use in a battery:

$\alpha \qquad \beta \qquad\qquad \delta \qquad\qquad \epsilon \qquad \zeta \qquad \eta$

Pb/ PbO/ KOH, H_2O/ PbO/ PbO_2/ Pb

The reactions are

$$PbO_2 + H_2O + 2e = PbO + 2OH^-$$
$$PbO + H_2O + 2e = Pb + 2OH^-$$

Derive the Nernst equation from the equilibrium relations and evaluate the standard cell potential.

3. The following cell has been considered for use in a battery with flowing electrolyte:

$\alpha \qquad\qquad \beta \qquad\qquad\qquad \delta \qquad\qquad \epsilon$

C/ $Ti^{3+}, Ti(OH)_2^{2+}$/ $FeCl_2, FeCl_2^+$/ C

where the reactions are

$$FeCl_2^+ + e = FeCl_2$$
$$Ti(OH)_2^{2+} + 2H^+ + e = Ti^{3+} + 2H_2O$$

Derive the Nernst equation from the equilibrium relations and evaluate the standard cell potential.

4. The following data are the activity coefficients of HCl and $CaCl_2$ at 25°C:

Molality	0.0001	0.0002	0.0005	0.001	0.002	0.005	0.01
HCl	0.989	0.984	0.975	0.966	0.952	0.928	0.904
$CaCl_2$	-	-	-	0.890	0.850	0.785	0.725

a) Compare the experimental values with those calculated from the Debye-Hückel limiting law.
b) Estimate the slope of the logarithm of the mean activity coefficient vs. the square root of ionic strength as the ionic strength approaches

zero. Compare this estimate with the predicted Debye-Hückel value.

5. Estimate the liquid-junction potential arising from

a) 0.001 N H_2SO_4 / 0.01 H_2SO_4

b) 0.001 N KCl / 0.01 N KCl

6. For the cell

α β δ ϵ

Fe/ 0.1 m $FeCl_2$/ 0.1 m $FeCl_3$, 0.1 m $FeCl_2$/ Pt

the reactions are

$$Fe^{2+} + 2e = Fe$$
$$Fe^{3+} + e = Fe^{2+}$$

a) Write the equilibrium expressions for the cell.

b) Calculate the value of the activity coefficients from Debye-Hückel theory for each component. Make reasonable assumptions to obtain numerical values.

c) Estimate the liquid-junction potential for this system.

d) Calculate the entire cell potential from the standard cell potential and from the corrections calculated above.

Bibliography

[1] R. A. Robinson and R. H. Stokes, *Electrolyte Solutions*, 2nd ed., (London: Butterworths Scientific Publications, 1959), Chap. 8.

[2] M. H. Lietzke, and R. W. Stoughton, *J. Phys. Chem.*, **66**, 508 (1962).

[3] L. A. Bromley, *AIChE J.*, **19**, 313 (1973).

[4] A. L. Horvath, *Handbook of Aqueous Electrolyte Solutions* (Chichester. England: Ellis Horwood Ltd., 1985), Sec. 2.8.

[5] J. F. Zemaitis, D. M. Clark, M. Rafel, and N. C. Scrivner, *Handbook of Aqueous Electrolyte Thermodynamics* (New York: AIChE, 1986).

[6] *The Measurement and Correction of Electrolyte Resistance in Electrochemical Tests*, L. L. Scribner and S. R. Taylor, eds., ASTM STP 1056, (Philadelphia: ASTM, 1990).

[7] W. H. Smyrl and J. Newman, *J. Phys. Chem.*, **72**, 4660 (1968).

[8] D. A. MacInnes, *The Principles of Electrochemistry* (New York: Reinhold Publishing Corp., 1939), p. 236.

[9] D. T. Sawyer and J. L. Roberts, *Experimental Electrochemistry for Chemists* (New York: John Wiley & Sons, 1974), p. 25.

Chapter 5

Electrode Kinetics

The rate at which a reaction at an electrode surface proceeds may be limited by the intrinsic kinetics of the heterogeneous process. In non-electrochemical systems, the macroscopically observed reaction rate is the result of a series of elementary processes, the details of which are rarely well-understood. A key difference between electrochemical and non-electrochemical processes is our ability to manipulate an additional driving-force variable: the electric potential. In industrial processes this ability to control the rate, through adjustment of the potential, affords enormous advantages in terms of process control. To put this ability in perspective, a change in potential of one volt at the surface of the electrode can cause a change in reaction rate of eight orders of magnitude; this change corresponds to an increase in temperature of several hundred degrees Celcius for a typical chemical reaction. In electrochemical systems we are also able to use traditional methods for controlling reaction rate and selectivity, e.g., temperature control and catalysis. An additional factor, which may be rate-limiting, is mass transfer. For the remainder of this chapter, we will assume that the supply of reactants to the electrode surface is sufficient to avoid mass transfer limitations. In the next chapter we will address these issues.

5.1   Role of the Interface

Electrode kinetics are governed by the potential difference across the thin (on the order of 10 Å) layer immediately adjacent to the electrode surface. Because of the unique role of this region, called the double layer, it has been studied in some detail in conjunction with electrode kinetics. Because the potential difference across this thin layer is on the order of 0.1 V in a typical reaction, the magnitude of the electric field is enormous (on the order of 10^6 V/cm). This large field provides an effective driving force for the electrode reaction. This region also provides an environment for charge separation leading to the reaction of a single ion. A local deviation from electroneutrality is analogous to the situation encountered near a central ion in the Debye-Hückel model.

In the thermodynamic analyses of the previous chapters, we considered equilibrium systems. Now we will analyze an electrode where the passage of a net current takes place. Passage of current implies that a reaction is occurring and that the system is no longer at equilibrium. A deviation of the electrode potential from the equilibrium condition is termed overpotential. An overpotential resulting from kinetic limitations is called an activation or surface overpotential. Because overpotential represents an energy loss and is related to the rate of heat production, a quantitative model of current density-overpotential behavior is required for engineering calculations. A fundamental model would proceed from a knowledge of elementary reactions, but few reactions have been studied in sufficient detail to develop a model in this way. More frequently, the equation form, derived from a fundamental model is used, but the parameters are determined from a fit of experimental data.

Electrode reactions are inherently heterogeneous in nature. This trait implies that a conductive surface must be placed in intimate contact with an electrolyte. Such an arrangement produces a number of problems when making measurements. Maintaining a clean, uniform electrode surface is a difficult task. Film formation, changes in electrode microstructure, and electrolyte contamination all contribute to variations in current-potential measurements over time. Measurements suitable for detailed studies require special steps such as careful electrode surface preparation, electrolyte purification, and control of mass transport to the electrode. In research studies the reproducibility of the electrode surface is sometimes maintained by using mercury, but this procedure is not feasible for industrial processes. Without proper precautions, reproducibility of current-potential curves to a precision better than 10% is difficult to achieve.

In the acquisition of current-potential data, measurement of the current is usually a straightforward process; placement of an ammeter in the external circuit is often sufficient. Accurate determination of the potential requires more effort. Because we cannot measure an absolute value of an electrode, we are forced to choose another electrode to provide a reference. Often a reference electrode is selected on the basis of characteristics such as reversibility, stability, and convenience. Another issue that must be addressed is suitable placement of the reference electrode to allow for correction of the ohmic difference between the working and reference electrode (ohmic compensation).

5.2 The Electric Double Layer

When we apply a potential to an electrode, the charges, which necessarily accumulate on the electrode surface, attract ions of the opposite charges from the electrolyte. If we consider a metal electrode and an aqueous electrolyte, then we expect to have the charges on the metal side balanced by an equal number of charges in a region adjacent to the electrode in the electrolyte. The simplest model of the electrolyte region is that of a line of charges at a fixed distance from the electrode surface. This arrangement, shown in Fig. 5.1, was first proposed by Helmholtz in 1879. The two layers of charge are then considered to comprise the double layer.

Because the field in the double layer is so large, solvent molecules like water, which possess a dipole moment, become oriented. The potential distribution in the double layer is complicated by this and other phenomena, but to a first approximation, we can model the double layer as a simple parallel-plate capacitor

$$C = \frac{\epsilon}{d} = \frac{D\epsilon_0}{d} \tag{5.1}$$

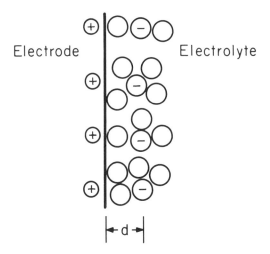

Figure 5.1: Model of the double layer proposed by Helmholtz. Two parallel layers of charge are separated by solvent molecules (unmarked circles) at a distance d, representing the outer Helmholtz plane.

where C is the capacitance per unit area, D is the dielectric constant (or relative permittivity), and d is the separation between the two layers of charge. If we take the distance between the layers of charge to be roughly the diameter of a hydrated ion, then d is on the order of 10 Å. Because of the high field, the dielectric constant of the water is lower than the bulk value ($D = 78$); measurements indicate that it is on the order of 10 in a field of this magnitude. From Eq. 5.1 the capacitance is

$$
\begin{aligned}
C &= \frac{(10)(8.9 \times 10^{-14} \text{ F/cm})}{10^{-7} \text{cm}} \\
&= 8.9 \times 10^{-6} \text{ F/cm}^2 \\
&\approx 10 \ \mu\text{F/cm}^2
\end{aligned}
\tag{5.2}
$$

This order of magnitude is correct for double-layer capacitance. From this model the potential distribution in the double layer is a simple linear function between the two layers of charge.

One test of the model is to measure the capacitance as a function of electrode potential. Classical measurements were performed

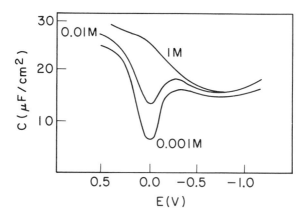

Figure 5.2: Capacitance vs. potential relative to the point of zero charge for a NaF solution on a mercury electrode at 25°C.

using dilute sodium fluoride solutions in contact with mercury electrodes. This system is appropriate for these measurements because NaF has little tendency for specific adsorption, i.e., adsorption due to chemical rather than electrical forces, and mercury offers a uniform surface. Results of capacitance vs. potential measurements [1] appear in Fig. 5.2. Although the capacitance is the right order of magnitude, it varies with potential.

More complex models were subsequently proposed. In 1910 Gouy and Chapman independently proposed their model of the electrode interface. It is the exact analog of the Debye-Hückel theory—but developed more than a decade before Debye-Hückel—applied to the planar geometry of the double layer. The thickness of the double layer represents a compromise between electrical forces tending to maintain the ordering in the region and thermal forces tending to make the arrangement more random. Significant deviations from electroneutrality occur in a region characterized by the Debye length.

$$\lambda = \left(\frac{\epsilon RT}{F^2 \sum_i z_i^2 c_{i,\infty}} \right)^{1/2} \tag{5.3}$$

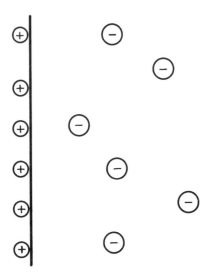

Figure 5.3: Gouy-Chapman model of the double layer. The net excess negative charge in the diffuse layer follows a Boltzmann distribution.

In this model ions are spread over an area near the electrode, rather than in a rigid plane (Fig. 5.3). As in the Debye-Hückel treatment of the previous chapter, we use Poisson's equation to describe the potential distribution, and the Boltzmann distribution to describe the distribution of ions.

$$\nabla^2 \phi = -\rho/\epsilon \tag{5.4}$$

$$c_i = c_{i\infty} \exp\left(\frac{-z_i F \phi}{RT}\right) \tag{5.5}$$

The charge density can be related to the concentration of ions by

$$\rho = F \sum_i z_i c_i \tag{5.6}$$

The combination of Eqs. 5.4, 5.5, and 5.6 yields

$$\nabla^2 \phi = -\frac{F}{\epsilon} \sum_i z_i c_{i\infty} \left(\frac{-z_i F \phi}{RT}\right) \tag{5.7}$$

In contrast to the Debye-Hückel treatment, we cast the Poisson-Boltzmann equation in Cartesian coordinates rather than in spherical coordinates.

$$\frac{d^2\phi}{dx^2} = -\frac{F}{\epsilon} \sum_i z_i c_{i\infty} \exp\left(\frac{-z_i F\phi}{RT}\right) \tag{5.8}$$

To derive analytical results, we consider the special case of a single symmetric electrolyte:

$$|z_+| = |z_-| = z \tag{5.9}$$

and

$$c_+ = c_- = c \tag{5.10}$$

where the concentrations all refer to bulk values. Substituting Eqs. 5.9 and 5.10 into Eq. 5.8 gives

$$\frac{d^2\phi}{dx^2} = -\frac{F}{\epsilon}\left[zc\exp\left(\frac{-zF\phi}{RT}\right) - zc\exp\left(\frac{zF\phi}{RT}\right)\right] \tag{5.11}$$

$$= \frac{2Fzc}{\epsilon} \sinh\left(\frac{zF\phi}{RT}\right) \tag{5.12}$$

The problem is further simplified by casting the potential and distance in dimensionless form:

$$\psi = \frac{zF\phi}{2RT} \tag{5.13}$$

$$\chi = \frac{x}{\lambda} \tag{5.14}$$

The Debye length λ for a symmetric electrolyte is

$$\lambda = \left(\frac{\epsilon RT}{F^2 \sum_i z_i^2 c_{i\infty}}\right)^{1/2} \tag{5.15}$$

$$= \left(\frac{\epsilon RT}{2F^2 z^2 c}\right)^{1/2} \tag{5.16}$$

Substitution of the dimensionless variables into Eq. 5.12 yields

$$\frac{d^2\psi}{d\chi^2} = \frac{1}{2}\sinh 2\psi \tag{5.17}$$

In terms of the dimensionless variables, the boundary conditions are

$$BC\ 1:\ \ \chi\ =\ 0 \qquad \psi\ =\ \psi_0 \qquad\qquad (5.18)$$

$$BC\ 2:\ \ \chi\ \to\ \infty \qquad \psi\ =\ 0 \qquad\qquad (5.19)$$

The first boundary condition implies that the potential at the metal-solution interface on the solution side is a fixed value. Later we will calculate ψ_0 in terms of the charge in solution from Gauss's law. The second boundary condition reflects the fact that far from a positively charged surface, the potential falls to zero. Eq. 5.17 can be solved by reduction of order. We define an electric field in terms of the following dimensionless variables:

$$E_f\ =\ -\frac{d\psi}{d\chi} \qquad\qquad (5.20)$$

The second-order derivative can be reduced to a first-order term.

$$\frac{d^2\psi}{d\chi^2}\ =\ -\frac{dE_f}{d\chi}\ =\ =\ -\frac{dE_f}{d\psi}\frac{d\psi}{d\chi}\ =\ E_f\frac{dE_f}{d\psi} \qquad\qquad (5.21)$$

When we substitute Eq. 5.21 into Eq. 5.17 and integrate, we obtain

$$E_f^2\ =\ \frac{1}{2}\cosh 2\psi\ +\ C_1 \qquad\qquad (5.22)$$

When we apply boundary condition 2, we can evaluate the integration constant C_1, and Eq. 5.22 becomes

$$E_f^2\ =\ \frac{1}{2}(\cosh 2\psi\ -\ 1)$$

$$=\ \sinh^2 \psi \qquad\qquad (5.23)$$

Substituting Eq. 5.20 into 5.23 and taking the square root yields

$$\frac{d\psi}{d\chi}\ =\ -\ \sinh \psi \qquad\qquad (5.24)$$

Integration of this equation yields

$$\ln[\tanh(\psi/2)]\ =\ -\chi\ +\ C_2 \qquad\qquad (5.25)$$

From the first boundary condition we obtain

$$\chi = \ln \left[\frac{\tanh(\psi_0/2)}{\tanh(\psi/2)} \right] \tag{5.26}$$

or

$$\psi = \ln \left[\frac{1 + \tanh(\psi_0/2) \exp(-\chi)}{1 - \tanh(\psi_0/2) \exp(-\chi)} \right] \tag{5.27}$$

From this expression it is difficult to discern the functionality of the potential distribution in solution. It is essentially an exponential decay; this functionality can be demonstrated by linearizing the exponentials in the hyperbolic sine term in Eq. 5.17. The resulting equation can be solved directly (see Problem 1).

From Gauss's law we can relate the potential gradient to the net charge in the double layer. Consider a control volume that has a face of unit area, e.g., 1 cm², at $x = 0$. A rectangular parallelepiped having such a face and extending out far enough into the solution will enclose all of the charge q (C/cm²). Gauss's law states that the integral of the outward normal component of the potential gradient is equal to the charge enclosed. Because we are considering a 1 cm² surface area, the charge enclosed is numerically equal to q, and at $x = 0$ we obtain

$$\epsilon \frac{d\phi}{dx} = q \tag{5.28}$$

We have an expression for the potential gradient from Eq. 5.24. When we substitute that expression into Eq. 5.28 and convert the dimensionless variables back to the original variables through Eqs. 5.13 and 5.14, we find

$$q = -(2RTc\epsilon)^{1/2} \sinh \left(\frac{zF\phi_0}{2RT} \right) \tag{5.29}$$

This equation relates the charge to the potential at the interface. For some value of ϕ_0 the right-hand side of the equation is zero; this value is referred to as the potential of zero charge.

Double-layer capacity is the variation of charge as a function of potential. The Helmholtz model is that of a parallel plate capacitor, and its capacity is invariant. In the Gouy-Chapman model the charge is a function of potential and concentration.

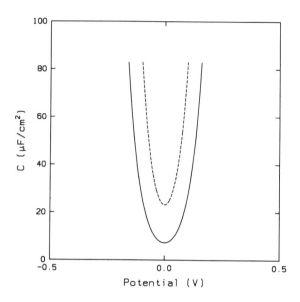

Figure 5.4: Capacitance of the double layer vs. potential calculated from the Gouy-Chapman model. The dashed curve represents a higher electrolyte concentration.

The magnitude of double-layer capacity is given by

$$C_d = \left(\frac{\partial q}{\partial \phi}\right)_{c,T} \tag{5.30}$$

$$= \frac{\epsilon}{\lambda} \cosh\left(\frac{zF\phi_0}{2RT}\right) \tag{5.31}$$

where the subscript c implies constant composition. From this equation we expect capacity to vary with the hyperbolic cosine of the potential. Over a small range of potential, the capacity based on the Gouy-Chapman model bears a resemblance to the behavior of actual systems. For example, a plot of Eq. 5.31 shows that the double-layer capacitance demonstrates the type of behavior for NaF electrolyte near the point of zero charge (Fig. 5.4). At potentials more than a half-volt in either direction, the observed flattening of the capacitance is not predicted from the theory.

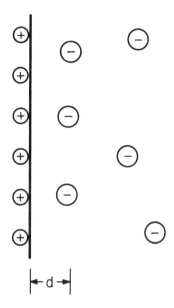

Figure 5.5: Stern model of the double layer. Some of the charge is fixed and some is spread out in the diffuse layer.

It is clear that neither the Helmholtz nor the Gouy-Chapman models give quantitative agreement over a large potential range. The former model is that of charges held firmly in a plane and the latter is that of charges spread out in a diffuse layer. Stern combined these ideas and developed a model of the double layer that included some of the charge in a fixed layer and some of the charge in a diffuse layer as shown in Fig. 5.5. Because these layers are in series, their capacitances should also behave as series capacitors, as such:

$$\frac{1}{C_s} = \frac{1}{C_h} + \frac{1}{C_{gc}} \tag{5.32}$$

where the subscripts s, h, and gc indicate Stern, Helmholtz, and Gouy-Chapman, respectively. From this equation we see that it is the smaller capacitance that governs the overall behavior of the system. If C_h is much greater than C_{gc}, then the overall capacitance is approximately equal to C_{gc}.

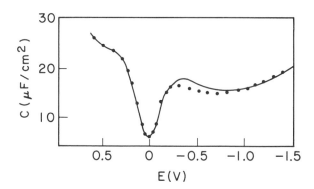

Figure 5.6: Capacitance of 0.001 M NaF solution as a function of potential. The solid line is calculated from the Stern model and the circles represent data.

There are several techniques for measuring double-layer capacity. These techniques include electrocapillary, ac impedance, and dropping mercury electrode methods. Using these techniques gives consistent results. We find that in concentrated solutions the capacitance of the diffuse part of the double layer is much greater than that of the Helmholtz portion; consequently, the smaller Helmholtz capacitance governs the overall behavior such that $C_s \approx C_h$. In Fig. 5.2 we see that capacitance is only approximately constant in concentrated solution. As concentration is reduced, the minimum in the curve becomes more pronounced, and the capacitance in the Helmholtz layer increases. Under these circumstances the overall capacitance is governed by the capacitance in the diffuse layer.

To test the Stern model, the Helmholtz capacity, measured at a high concentration, can be taken as a constant over the entire concentration range. Addition of that contribution to the Gouy-Chapman capacity gives a curve for the capacitance over a range of potential. As shown in Fig. 5.6 the agreement between calculated and experimental results [1] lends support to this model.

An important consequence of the double-layer structure is that species outside the Helmholtz region are too distant to react. This separation implies that the driving force for the reaction is the potential drop across the Helmholtz region, rather than the potential drop across the entire double layer. Moreover, the effective concentration driving the reaction is different from the bulk concentration. The late Soviet electrochemist Alexander Frumkin carried out an extensive series of investigations to confirm this result, and the correction for the concentration variation in the double layer is called the Frumkin correction.

When studying electrode kinetics, it is desirable to eliminate factors that interfere with making a measurement of the intrinsic electrode kinetics. To minimize double-layer effects, an investigator can add a non-reacting (supporting) electrolyte to the solution. This addition has the effect of increasing the capacitance of the diffuse layer; consequently, the overall capacitance is approximately equal to the capacitance of the Helmholtz layer.

5.3   An Electrode Kinetics Model

In ordinary chemical kinetics we frequently depict the progress of a reaction by plotting the reaction coordinate against the energy of the species as shown in Fig. 5.7. Electrochemical reactions are treated analogously, but because charge transfer is involved, we must include the effects of potential in the energy.

If we consider the reduction of an ion progressing at potential ϕ_1 then we have a picture similar to that of the chemical reaction. We will consider the general reduction of a species O to the product R.

$$O^+ + e = R \tag{5.33}$$

We know that a more negative potential tends to promote reduction, but by our convention a more negative potential corresponds to a more positive energy. At progressively more negative potentials ϕ_2 and ϕ_3, the energy of the oxidized species is increased. If we take the metal potential to be fixed, then the energy of the reduced species remains constant, and the energy vs. reaction coordinate can be drawn as in Fig 5.8. The potential ϕ_1 is the most positive potential and corresponds to a potential where oxidation is favored. At the lowest potential ϕ_3 the reduced form R is favored. An intermediate potential ϕ_2 is near the equilibrium potential where no net reaction takes place.

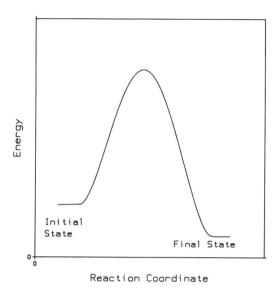

Figure 5.7: Energy along a reaction coordinate for an ordinary chemical reaction.

To develop a quantitative description of the energetics of a reaction, consider the case where we begin an experiment at the potential ϕ_1 and reduce it to ϕ_2 as shown in Fig. 5.8. The activation energy for the first process is G_{c1}. It is a lower value G_{c2} at ϕ_2. Although we have increased the energy by the amount $nF(\phi_2 - \phi_1)$, only a fraction of this amount has been effective in lowering the activation barrier. We denote this fraction as β, the symmetry factor. The energy barrier is changed by the amount $\beta nF(\phi_2 - \phi_1)$

$$G_{c2} = G_{c1} + \beta nF(\phi_2 - \phi_1) \tag{5.34}$$

where the subscript c indicates a cathodic process. Note that ϕ_2 is more negative than ϕ_1, which implies that the second term on the right-hand side makes a negative contribution to the activation energy. The activation energy in the anodic direction is increased by the amount $(1 - \beta)nF(\phi_2 - \phi_1)$.

$$G_{a2} = G_{a1} - (1 - \beta)nF(\phi_2 - \phi_1) \tag{5.35}$$

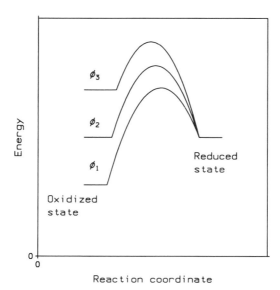

Figure 5.8: Energy along a reaction coordinate for an electrochemical reaction at three different potentials, $\phi_1 > \phi_2 > \phi_3$.

The symmetry factor is a number that can vary between 0 and 1. It is the fraction of the potential across the double layer promoting the cathodic reaction. The experimental values are generally close to $1/2$. Although we have retained the symbol n, it is unlikely that more than one electron participates in any elementary step.

The significance of the symmetry factor becomes clearer if we look at the two limiting cases. If $\beta = 0$, then the situation depicted in Fig. 5.9a results. None of the energy has promoted the cathodic reaction, and $G_{c1} = G_{c2}$. At the other extreme, $\beta = 1$, and all of the additional energy goes into promoting the cathodic reaction, as shown in Fig. 5.9b.

We now have an expression for activation energy as a function of potential. The form of our kinetic expression is the same as that for chemical reactions

$$k = k' \exp\left(\frac{-G^{\ddagger}}{RT}\right) \tag{5.36}$$

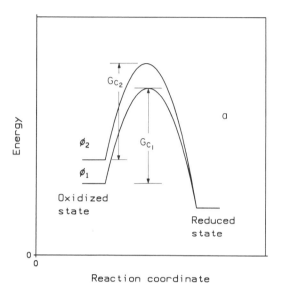

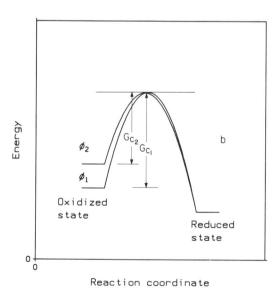

Figure 5.9: a) Potential energy diagram for $\beta = 0$; b) Diagram for $\beta = 1$.

where $G^{\ddagger}$ is the free energy of activation and k' is a constant. For the cathodic and anodic portions of our reactions, the activation energies are given by Eqs. 5.34 and 5.35, respectively. The rate of an electrochemical reaction is directly proportional to the current density, as shown below

$$r = \frac{i}{nF} = k'c\exp\left(\frac{-G^{\ddagger}}{RT}\right) \tag{5.37}$$

where r is the reaction rate (mol/s cm^2), and c is the reactant concentration (mol/cm^3). For a general anodic reaction involving a single electron transfer, the rate can be expressed in terms of the potential by substituting Eq. 5.34 into Eq. 5.36. If we choose a reference potential, we can drop the subscripts.

$$r_a = \frac{i_a}{nF} = k'_a c_R \exp\left\{-\frac{[G_a - (1-\beta)nF\phi]}{RT}\right\} \tag{5.38}$$

We can absorb the constant G_a, which corresponds to the activation energy at our reference potential, into the rate constant:

$$r_a = k_a c_R \exp\left[\frac{(1-\beta)nF\phi}{RT}\right] \tag{5.39}$$

For a first-order reaction k_a has the units of cm/s; for a reaction of order p, the rate constant is in units of $(cm/s)(cm^3/mol)^{p-1}$. Similarly, the rate equation in the cathodic direction is

$$r_c = \frac{i_c}{nF} = k_c c_O \exp\left(\frac{-\beta nF\phi}{RT}\right) \tag{5.40}$$

The net anodic current density is the difference between the magnitudes of the anodic and cathodic current densities.

$$
\begin{aligned}
r &= r_a - r_c \\
&= \frac{i_a - i_c}{nF} \\
&= \frac{i}{nF} \\
&= k_a c_R \exp\left[\frac{(1-\beta)nF\phi}{RT}\right] - k_c c_O \exp\left(\frac{-\beta nF\phi}{RT}\right)
\end{aligned} \tag{5.41}
$$

At equilibrium the net current density is zero, but the rates of the anodic and cathodic reactions are not zero. The magnitude of this (equal and opposite) current density is called the exchange current density i_0. If we designate the equilibrium potential as ϕ^o, then at equilibrium the magnitude of i_0 is

$$\frac{i_0}{nF} \;=\; k_a c_R \exp\left[\frac{(1-\beta)nF\phi^0}{RT}\right] \tag{5.42}$$

$$=\; k_c c_O \exp\left(\frac{-\beta nF\phi^0}{RT}\right) \tag{5.43}$$

The net current at equilibrium is zero, and Eq. 5.41 becomes

$$\frac{i}{nF} \;=\; 0 \;=\; k_a c_R \exp\left[\frac{(1-\beta)nF\phi^0}{RT}\right]$$
$$-\; k_c c_O \exp\left(\frac{-\beta nF\phi^0}{RT}\right) \tag{5.44}$$

When we take the logarithm of this equation, we obtain

$$\ln(k_a c_R) \;+\; \frac{(1-\beta)nF\phi^0}{RT} \;=\; \ln(k_c c_R) \;-\; \frac{\beta nF\phi^0}{RT} \tag{5.45}$$

By rearranging, this equation can be expressed as

$$\phi^0 \;=\; \frac{RT}{nF}\ln\left(\frac{k_c}{k_a}\right) \;-\; \frac{RT}{nF}\ln\left(\frac{c_R}{c_O}\right) \tag{5.46}$$

This equation gives the reversible potential at a specified concentration. This function is the same that the Nernst equation serves, and the terms of this equation can be identified with corresponding terms in the Nernst equation. Note that the term we have designated ϕ^0 is concentration-dependent and corresponds to E. The first term on the right-hand side is concentration-independent and corresponds to E^o.

The overpotential represents a departure from the equilibrium potential and can be defined as

$$\eta_s \;=\; \phi \;-\; \phi^0 \tag{5.47}$$

If we substitute Eqs. 5.46 and 5.47 into Eq. 5.41 we obtain

$$i = nFk_a c_R \exp\left[\frac{(1-\beta)nF}{RT}\left(\eta_s + \frac{RT}{nF}\ln\frac{k_c}{k_a} + \frac{RT}{nF}\ln\frac{c_O}{c_R}\right)\right]$$

$$-nFk_c c_O \exp\left[\frac{-\beta nF}{RT}\left(\eta_s + \frac{RT}{nF}\ln\frac{k_c}{k_a} + \frac{RT}{nF}\ln\frac{c_O}{c_R}\right)\right] \quad (5.48)$$

Rearranging gives us

$$i = nFk_c^{1-\beta}k_a^{-\beta}c_O^{1-\beta}c_R^{\beta}$$
$$\left[\exp\frac{(1-\beta)nF}{RT}\eta_s - \exp\frac{-\beta nF}{RT}\eta_s\right] \quad (5.49)$$

At equilibrium the overpotential is zero, and each component of the net current density must be equal to the exchange current density; therefore, we can identify the term outside the brackets with i_0. We also define the transfer coefficients for the anodic and cathodic components as

$$\alpha_a = (1-\beta)n \quad (5.50)$$
$$\alpha_c = \beta n \quad (5.51)$$

The terminology regarding the transfer coefficient and symmetry factor varies widely. Some authors prefer to drop the n in the definition of the transfer coefficient and include it explicitly in the kinetic equation as αn. With the definitions in Eqs. 5.50 and 5.51, $\alpha_a + \alpha_c = n$. Equation 5.49 can be expressed in terms of the exchange current density and transfer coefficients, as

$$i = i_0\left[\exp\frac{\alpha_a F}{RT}\eta_s - \exp\frac{-\alpha_c F}{RT}\eta_s\right] \quad (5.52)$$

This expression is the Butler-Volmer equation. It is a standard model that can be used to describe the current-overpotential relationship for an electrode at a specified temperature, pressure, and concentration of reacting species. A plot of the current-overpotential relation is commonly referred to as a polarization curve. Three variables, α_a, α_c, and i_0, must be determined in this kinetic equation. For certain reactions, the Butler-Volmer equation provides a reasonable description of current-overpotential behavior, and the variables are determined from experimental data. Even when the reaction mechanism is known, it may not be possible to cast the overall current-overpotential relation in Butler-Volmer form; in those cases an alternate specialized equation can often be developed. In most cases the detailed mechanism is not known, but the Butler-Volmer equation—or a simplified form of it—can be used as a fitting function that is valid over a restricted range.

Frequently, several reactions occur simultaneously on the same electrode surface. For example, it is possible to deposit copper and evolve hydrogen on the same copper cathode. If we are operating in a range where each electrode reaction is kinetically controlled, then we can use an appropriate polarization expression for each reaction and express the total current density as the sum of the currents for each individual reaction. If we consider two reactions both described by Butler-Volmer kinetics, then we will need to determine the three parameters (α_a, α_c, and i_0) for each reaction; in general, these values will be different for each reaction. The overpotential for each reaction must be calculated or measured with respect to its own equilibrium potential in accordance with Eq. 5.47.

The exchange current density is a function of the concentration of both reactants and products. If one of the charge transfer steps is rate-limiting, then we expect the exchange current density to vary approximately as the square root of concentration as indicated in Eq. 5.49; in this analysis we assume that the symmetry factor is one-half. The material used for the electrode strongly influences the reaction rate. For example, the exchange current density can vary by five orders of magnitude or more for the same reaction on different types of electrodes.

Exchange current density is sensitive to contamination of the electrode surface. Measured values can change by an order of magnitude when contaminants are present in the electrolyte. A large value of exchange current density is characteristic of an electrode reaction that is said to be "reversible." In other words, an electrode reaction proceeds with little overpotential at high current density. The term reversible is relative but often refers to a reaction where the exchange current density in aqueous electrolyte is greater than 1 mA/cm^2. At the oppo-

Electrode	Couple	Electrolyte	$\log i_0$ (A/cm^2)
Pt	Fe^{3+}/Fe^{2+}	H_2SO_4	-3
Pt	H^+/H_2	H_2SO_4	-3
Hg	H^+/H_2	H_2SO_4	-11
Cu	Cu^{2+}/Cu	$CuSO_4/H_2SO_4$	-3
Pt	H_2O/O_2	$HClO_4$	-10

Table 5.1: Exchange current densities for electrochemical reactions at 25°C.

site extreme, electrode reactions with low exchange current densities are referred to as sluggish or "irreversible." When a sluggish reaction occurs at an electrode, the electrode is said to be polarizable. The transfer coefficients are usually taken to be independent of concentration and temperature. Transfer coefficients typically vary between 0.2 and 2. Values of exchange current densities for several common reactions appear in Table 5.1.

In general, metal deposition reactions have relatively high exchange current densities. The oxygen reduction reaction has a very low exchange current density on all electrodes; this impediment is significant to the development of low-temperature energy conversion devices. Most non-metal reactions involving the transfer of more than two electrons tend to be sluggish. Because the oxidation of the simplest hydrocarbons requires the transfer of multiple electrons (8 electrons for methane), these reactions are also sluggish. For this reason most low-temperature fuel cells require hydrogen at the anode.

The exchange current density is a strong function of temperature. Several fuel cells currently under development operate at temperatures in the range of 600 - 1000°C, where the exchange current densities for hydrogen oxidation and oxygen reduction are high, and the surface overpotentials are small.

A typical plot of overpotential vs. current density, calculated from Eq. 5.52, appears in Fig. 5.10. The curve is symmetric about the origin only for the case where $\alpha_a = \alpha_c$. In the case of the copper reaction, $\alpha_c = 0.5$, $\alpha_a = 1.5$, and $i_0 = 0.001$ A/cm². These values are

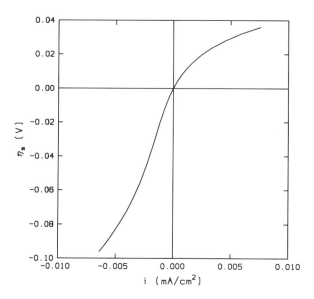

Figure 5.10: Overpotential vs. current density (polarization curve) for the copper dissolution/deposition reaction.

valid for the copper dissolution/deposition reaction in acid solution near room temperature. The Butler-Volmer equation then becomes

$$i \; = \; 0.001(\exp \; 58.1\,\eta_s \; - \; \exp \; -19.4\,\eta_s) \qquad (5.53)$$

By our convention, anodic currents and overpotentials are positive, and cathodic currents and overpotentials negative.

5.4 Simplified Electrode Kinetics Models

The Butler-Volmer expression is somewhat inconvenient because we cannot solve explicitly for overpotential. In the special case where the transfer coefficients are equal, the exponentials can be cast in the form of a hyperbolic sine function. For the case of equal transfer coefficients, Eq. 5.52 becomes

$$i = 2i_0 \sinh \frac{\alpha F}{RT} \eta_s \tag{5.54}$$

When the magnitude of the overpotential is large, another simplification is possible. If the overpotential is large and positive, then the first exponential term in Eq. 5.52 is much larger than the second term, and the polarization expression becomes

$$i = i_0 \exp \frac{\alpha_a F}{RT} \eta_s \tag{5.55}$$

This expression is the Tafel equation. Because it can be written explicitly in terms of current density or overpotential, it is commonly used in modeling electrochemical processes. When the overpotential is a large negative value, the first term in Eq. 5.52 can be neglected, and an expression analogous to Eq. 5.55 can be written.

The Tafel equation can be rearranged in the following several equivalent forms:

$$\eta_s = \frac{RT}{\alpha_a F} \ln \frac{|i|}{i_0} \tag{5.56}$$

$$= \frac{2.303 RT}{\alpha_a F} \log \frac{|i|}{i_0} \tag{5.57}$$

$$= B \log |i| - A \tag{5.58}$$

where

$$B = \frac{2.303 RT}{\alpha_a F} \tag{5.59}$$

$$A = \frac{2.303 RT}{\alpha_a F} \log i_0 \tag{5.60}$$

The constant B is called the Tafel slope. Use of the Tafel approximation depends on the error that can be tolerated; it is generally used when the overpotential is at least 50 to 100 mV.

When the current density increases by a factor of 10, the overpotential increases by B volts. The constant 2.303 RT/F is approximately 0.059 V. Because α is usually between 0.2 and 2, the Tafel slope typically varies between 30 and 300 mV/decade. The values for the Tafel slope and exchange current density often represent an empirical fit of the data. Because current density enters into the logarithmic term, we cannot maintain our convention of a negative value for a cathodic current. Instead, we use the absolute value of the current density and apply the proper sign for the overpotential.

One method for determining the Tafel parameters is to plot the logarithm of the current density vs. the overpotential, and analyze the slope and intercept of the curve based on high current density behavior. In Fig. 5.11, Eq. 5.53 is plotted in a semi-logarithmic format. The dashed line is an extrapolation of high current density data to the line where the overpotential is zero. If the kinetic equation follows the Butler-Volmer form, then extrapolation of either the anodic or cathodic data to $\eta_s = 0$ gives the same result. In this case we know that the exchange current density is 1 mA/cm². In general, we would make this type of plot from experimental data, and we would determine i_0 from the plot. The slope of the line gives the Tafel slope, and the intercept on this semi-log plot gives the exchange current density. In making such a measurement, it is important to eliminate—or take into account—non-kinetic effects, e.g., mass transport limitations.

Inspection of Eq. 5.56 reveals that there is an explicit proportionality between overpotential and temperature. On initial analysis, we might expect the overpotential to increase with increasing temperature. Such an expectation is counter-intuitive. The overpotential represents a "kinetic resistance," but we would expect increasing temperature to promote, rather than retard, an electrode reaction. Measurements [2] reveal that the overpotential, in fact, decreases with increasing temperature at a specified current density (Fig. 5.12). This trend can be attributed to the more rapid increase in the values of the exchange current density with temperature.

A second approximate form can be developed for the low overpotential region. When η_s is small, the exponential terms in Eq. 5.52 can be expanded in a Maclaurin series as such

$$\begin{aligned}
i &= i_0 \left[1 + \frac{\alpha_a F}{RT} \eta_s - 1 + \frac{\alpha_c F}{RT} \eta_s \right] \\
&= \frac{i_0(\alpha_a + \alpha_c)F}{RT} \eta_s
\end{aligned} \tag{5.61}$$

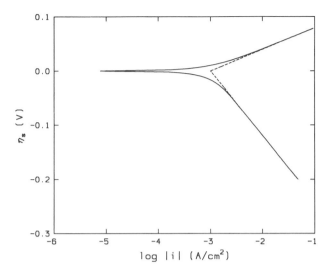

Figure 5.11: Use of Tafel extrapolation to determine the exchange current density. The dashed line represents extrapolation from high current density data.

This expression is the linear form of the Butler-Volmer equation. It has the advantage that it can be solved explicitly for the overpotential. Its use as a single parameter equation is common in modeling low current density processes. Although the accuracy of this approximation varies with the values of the parameters, it is often used when the overpotential is 10 mV or less. In addition, if the current density does not vary widely ($\pm$ 30%), an empirical, linear relation can be used, even in the high current density region.

5.5 Reference Electrodes

Measurement of overall cell potential at equilibrium or under load is a simple matter. For typical reactions one can use a voltmeter, and measure the potential difference across the electrodes. If there is a single reaction at each electrode, at equilibrium this measurement may be sufficient to give a good estimate of the reversible potential

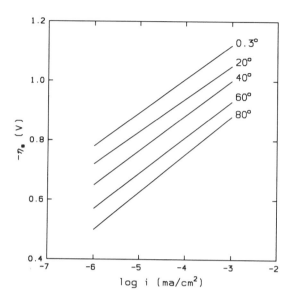

Figure 5.12: Overpotential vs. logarithm of current density as a function of temperature for the hydrogen evolution reaction on mercury.

of a working electrode. Such a determination depends on the use of a counterelectrode having a known reversible potential. It is important to use a potential-measuring device that only permits a small current flow so that significant overpotentials are avoided. Many inexpensive voltmeters have an input impedance of at least 10 megohms, which is sufficient for most engineering measurements. Greater input impedance ($> 10^8$ megohms) is available from an electrometer. It is only those reactions with small exchange current densities where more sensitive instrumentation is required for the determination of reversible potential.

Once a cell is displaced from equilibrium either through spontaneous discharge or through imposition of an applied potential, then measurement of the potential of one electrode with respect to a reference becomes more complicated. Under load, the overall cell potential is comprised of several components including the reversible cell potential, the overpotential at each electrode, and ohmic drop between the electrodes. A simple two-electrode system that was satisfactory for making equilibrium measurements is no longer adequate for accurate measurements under load. Attempting to disentangle the components from a two-electrode measurement is usually difficult and sometimes impossible.

A common technique for overcoming this problem is to insert a third electrode into the electrolyte as shown in Fig. 4.7. An electrode suitable for determining the relative potential of one of the electrodes is termed a reference electrode. Such an electrode is usually operated so that it draws essentially no current, and it can be moved to different positions in the electrolyte. The fact that it draws only a miniscule current implies that the overpotential at the reference electrode is negligible, and all of the overpotential that is measured can be attributed to the working electrode. Because of the mobility of the reference electrode, we can sense a local overpotential, even when the current distribution is nonuniform.

Although we would generally like to use reference electrodes that operate at a known, reversible potential, there are exceptional cases. In some corrosion measurements, where extreme accuracy is not required, a corroding zinc electrode can be used as a reference electrode. In other cases the electrolyte environment is unsuitable for a standard commercial reference electrode, and another reference must be found. For example, at high temperatures simple metal electrodes are sometimes used as reference electrodes. Although electrochemical reactions at the reference electrode are poorly characterized, the electrode may give a stable reading. Such electrodes are termed pseudo-reference electrodes.

A reference electrode is commonly used to measure the overpotential at a working electrode. Because the overpotential represents a voltage loss, we are often interested in analyzing a cell to determine the sources of the loss. Adding a second reference electrode can be used to determine the potential drop between two points in solution. This setup is sometimes called a four electrode arrangement for measuring ohmic drop. Although there are several operational difficulties, which are addressed below, we can often obtain a good estimate of all components of potential difference in a cell through the use of reference electrodes.

Another use of reference electrode measurements is to determine the current distribution in a cell having nonuniform current distribution. Direct measurement of local current densities is difficult because the point of origin for any current entering (or leaving) an electrode/electrolyte interface cannot easily be established once current resides within the electrode. It is possible to make a polarization curve for the system of interest in a well-characterized laboratory cell. This curve then serves as a "calibration curve" for determining local current density when the overpotential is measured at different points along an electrode surface.

Overpotential is often classified either as surface overpotential, associated with sluggish electrode kinetics, or as concentration overpotential, associated with mass transfer limitations. Several other terms are sometimes used in connection with voltage losses, but their meanings are ambiguous. Ohmic overpotential is sometimes used to denote ohmic drop in solution but in other contexts it means voltage drop through an electrode film. The term "depolarizer" often denotes a different reactant that reduces the reversible potential of an electrolytic cell. In this context the reduction in potential is due to a thermodynamic rather than kinetic phenomenon.

Overpotential and ohmic drop always represent voltage—and energy—losses. If we maintain the sign conventions established in Ch. 2, then we can express the total voltage loss in a cell as

$$E_{loss} \ = \ \eta_{s,a} \ + \ \eta_{cn,a} \ - \ \eta_{s,c} \ - \ \eta_{cn,c} \ + \ \Delta\phi_{ohm} \qquad (5.62)$$

where the subscripts denote: s for surface, a for anodic, cn for concentration, c for cathodic, and ohm for ohmic. The cathodic components enter with a negative sign because the values of cathodic overpotentials are negative, and each term makes a positive contribution to the total. The sign of each overpotential component can be computed directly from the definition of overpotential (Eq. 5.47). Calculation of overpotential from reference electrode measurements is illustrated in the next section. Because each term represents a loss component, the absolute values are sometimes used, and the conventions regarding signs are avoided. Although this expression is fairly general there are other possible sources of voltage loss that might need to be included. For example, ohmic losses in the electrodes may be important in high current cells.

To relate overall cell potential to the loss component we need to consider whether we have an electrolytic (energy consuming) process or a galvanic (energy producing) cell. If we have an electrolytic cell operating reversibly, then the applied potential will equal the thermodynamic requirement E. In reality there are always irreversible losses E_{loss} associated with a cell passing a finite net current. The total potential that must be applied to the electrodes is the sum of the thermodynamic and loss components.

$$E_{appl} \ = \ E \ + \ E_{loss} \qquad \text{electrolytic} \qquad (5.63)$$

For a galvanic cell the losses are subtracted from the reversible potential.

$$E_{appl} \ = \ E \ - \ E_{loss} \qquad \text{galvanic} \qquad (5.64)$$

5.5.1 Potential Distribution in Solution

With appropriate reference electrodes we are in a position to estimate the magnitudes of the irreversible losses. To analyze the components of potential drop, consider a simple plane parallel electrode arrangement with a uniform current distribution under load. As an example, we will treat the Daniell cell. We can sketch the potential distribution under different conditions, and treat it for both the galvanic and electrolytic cases. For our analysis we will use a standard hydrogen electrode (SHE) as a reference electrode. If we are operating under standard conditions in the galvanic mode with a high resistance between the electrodes, then the cell will be essentially at equilibrium. In the galvanic mode the zinc electrode is the anode (negative electrode) and the copper electrode is the cathode. The potential distribution can be represented schematically as shown in Fig. 5.13a. At equilibrium there is no ohmic drop in solution, and the potential in the electrolyte is uniform. The potential difference across the double layer is indicated schematically by the rapid potential change near each electrode.

If the resistance of the load is decreased, more current flows through the circuit, and irreversible voltage losses occur. Let us first consider

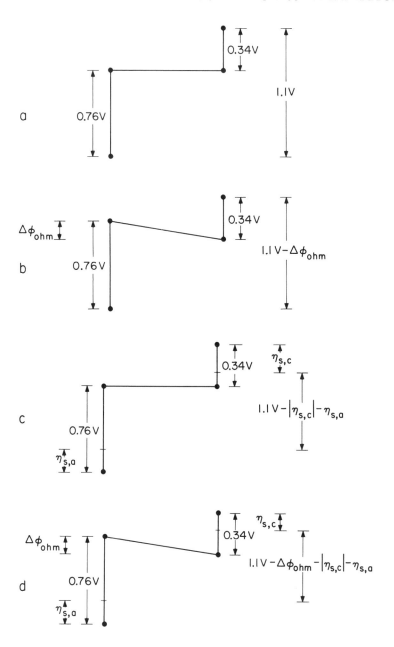

Figure 5.13: Potential distribution in solution for a galvanic cell with a uniform field: a) Equilibrium; b) Resistive electrolyte; c) Sluggish reactions; d) General case.

the case where the electrolyte resistance is very high, and the overpotential is negligible (Fig. 5.13b). There are several points to note. The potential difference across the electrodes is reduced. At equilibrium the potential difference is 0.34 V - (-0.76 V) = 1.1 V. In the galvanic cell the potential difference is reduced by an amount equal to the ohmic loss in the cell. Current in this cell flows from left to right. On initial inspection it may seem paradoxical that the current—composed of positive charges—flows toward the positive electrode, but from the diagram it is clear that the potential outside the double layer is lower on the right-hand side than on the left-hand side. This potential drop, between points just outside the double layer, can be measured directly by connecting a voltmeter between two reference electrodes, each positioned near one of the electrodes. Because we have assumed that the overpotential is negligible, the voltmeter reading between a reference electrode and an adjacent electrode is the same as in the equilibrium case.

Let us now consider a third case where the electrolyte is highly conductive and the electrode kinetics are sluggish (Fig. 5.13c). Because there is negligible potential drop in solution the potential distribution is represented by the horizontal line. By our convention, overpotential at the anode is a positive quantity, while that at the cathode is a negative quantity. Total cell potential is reduced by the algebraic sums of the overpotentials.

$$E_{loss} = \eta_{s,a} - \eta_{s,c} \tag{5.65}$$

We need to calculate an overpotential from a reference vs. working electrode measurement. Using the proper conventions will help us keep the signs straight. When making a measurement, we always connect the negative lead of the voltmeter to the reference electrode. The sign on the voltmeter reading is the one that we use in calculations. Overpotential represents a deviation from the equilibrium potential. At equilibrium the potential of the working vs. the reference electrode potential is

$$E_{wr,rev} = E_{w,rev} - E_{r,rev} \tag{5.66}$$

where w indicates the working electrode, r indicates the reference electrode, wr indicates working vs. reference and rev indicates reversible behavior. Under load, the potential measured on the voltmeter is

$$E_{meas} = E_w - E_{r,rev} \tag{5.67}$$

The overpotential is the difference between the measured and equilibrium reference vs. working electrode values; subtraction of Eq. 5.66 from Eq. 5.67 yields

$$\begin{aligned}
\eta &= E_{meas} - E_{wr,rev} \\
&= (E_w - E_{r,rev}) - (E_{w,rev} - E_{r,rev}) \\
&= E_w - E_{w,rev} \tag{5.68}
\end{aligned}$$

The more general case is one where both ohmic drop and overpotential are important components (Fig. 5.13d). It represents a straightforward extension of a combination of the two cases considered above. Total cell potential is reduced by the sum of the overpotentials and ohmic drop.

We can perform a similar analysis for an electrolytic cell. As an example we consider operating the Daniell cell in reverse. This operation can be accomplished by connecting the cell to another potential source having a potential larger than the reversible potential of the Daniell cell. In this case the negative electrode becomes the cathode and the positive electrode becomes the anode but, the signs of the electrode do not change. The potential diagram is shown in Fig. 5.14. Total cell potential is the sum of the losses plus the reversible potential. When the voltmeter is connected according to convention with negative lead on the reference electrode, the reading at the cathode is negative and the reading at the anode is positive.

5.5.2 Ohmic Compensation

In the above calculations we have assumed that we could place the reference electrode just outside the double layer and sense the potential at that point. Because of operational constraints, the reference electrode is placed much farther away from the double layer. Once the reference electrode is removed some distance from the working electrode, then the potential sensed is different from the potential that we want to measure. The difference in potential is due principally to the

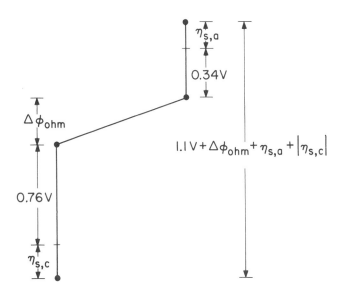

Figure 5.14: Potential in solution for an electrolytic cell.

ohmic potential difference between the point just outside the double layer and the point of measurement. In highly conductive electrolyte the ohmic drop may be negligible, as illustrated in Fig. 5.13c. In this case placing the reference electrode away from the working electrode makes little difference in the value that would be read on a voltmeter connected between the working and reference electrodes. In Fig. 5.13d we see that a reference electrode located away from the working electrode would include a large ohmic component in the measurement.

The calculation or measurement of the ohmic component to correct a reference electrode measurement is referred to as ohmic compensation. There are numerous techniques for making the correction. In a laboratory cell one might measure the conductivity and calculate the ohmic drop directly. This technique is easiest for a cell with a uniform current distribution. Alternatively, the total ohmic drop can be directly measured between two reference electrodes placed in solution.

In practice a reference electrode is placed from several millimeters to several centimeters away from the working electrode in a typical laboratory cell. Optimum reference electrode placement is usually a matter of several compromises. Placing a reference electrode close to the working electrode reduces total ohmic drop, but if it is too close it can shield the working electrode. This shielding effect reduces current flow to the working electrode which, in turn, reduces the overpotential; therefore, the magnitude of the overpotential resulting from a reference electrode measurement is lower than the overpotential that would result in the absence of a reference electrode. Shielding is readily avoided by moving the reference electrode farther away from the working electrode. A general rule is to move the reference electrode to a distance at least four reference electrode diameters from the working electrode. For a reference electrode at this distance, upstream disturbance caused by the reference electrode is small and the current essentially re-establishes itself on the working electrode.

There are several additional points to consider for optimum reference electrode placement. It is generally desirable to avoid locating the reference electrode in regions of high potential gradient, where small errors in electrode placement result in large errors in ohmic compensation. If an electrolyte is resistive, then we would like to place the reference electrode closer to the working electrode; otherwise, compensation for the ohmic term results in the subtraction of one large number from another with correspondingly large errors. One technique for reducing the diameter of the reference electrode to facilitate closer placement is through the use of a capillary tip (Luggin capillary) and placement of the electrode portion of the reference electrode in a remote location as shown in Fig. 5.15. With such an arrangement a thin probe (on the order of 1 mm) can be placed close to the working electrode with only a small disturbance of the field. Ohmic drop between the probe tip and the electrode portion of the reference electrode is negligible because the current flow in the connecting tube is extremely small.

5.5.3   Types of Reference Electrodes

Although the standard hydrogen electrode (SHE) is the primary reference, it has few other advantages to recommend it for engineering purposes. Because gas must be fed to it continuously, it cannot be made in a compact, self-contained package. And, because this elec-

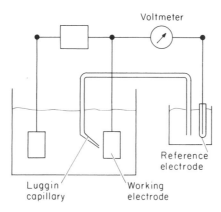

Figure 5.15: Schematic of the Luggin capillary arrangement. The reference electrode is in a remote location.

trode is unwieldy, a number of other reference electrodes have gained general acceptance and are frequently used in engineering measurements. Several are well-characterized and available commercially. A number of reference electrodes and appropriate applications are reviewed by Ives and Janz [3].

In choosing a reference electrode we are generally interested in factors such as reproducibility, stability, and small temperature sensitivity. Use of a reference electrode of the same type as the working electrode avoids problems with contamination and liquid-junction potentials. For example, a copper metal electrode placed directly in the electrolyte solution can serve as a convenient reference electrode in a copper/copper sulfate system. By contrast, it would be inappropriate to use an oxygen reference electrode because it is highly irreversible. Corroding electrodes are also unsuitable for work where high accuracy is required. To minimize liquid-junction potentials, concentrations of the reference cell electrolyte are closely matched with the electrolyte in the working electrode compartment when convenient.

One of the most common reference electrodes is the calomel electrode. It is constructed by covering a pool of mercury with mercurous chloride (calomel). Electrical contact is maintained with a platinum wire contacting the mercury. Potassium chloride is the electrolyte for the following electrode reaction:

$$Hg_2Cl_2 \ + \ 2e \ = \ 2Hg \ + \ 2Cl^- \tag{5.69}$$

By convention the activity of the solid phases is at unit activity. At 25°C the reversible potential is

$$\begin{aligned} E \ &= \ E^\circ \ - \ \frac{2.303RT}{2F} \ \log a_{Cl^-}^2 \\ &= \ 0.268 \ - \ 0.059 \log a_{Cl^-} \end{aligned} \tag{5.70}$$

Commercial electrodes are commonly prepared with three concentrations of KCl: 0.1 N, 1.0 N, and saturated; the corresponding reversible potentials are 0.334, 0.281, and 0.242 V, respectively. Literature values vary by one or two millivolts because of uncertainties connected with electrode aging and liquid-junction potentials. The electrode with 0.1 N KCl has the lowest temperature coefficient, -0.06 mV/°C, while the saturated calomel electrode (SCE) has a temperature coefficient of -0.65 mV/°C. By convention the SHE is assigned a potential of 0.00 V at any temperature, and the temperature coefficient actually contains contributions from both electrodes; however, the temperature coefficient of the SHE is arbitrarily taken as zero.

The silver-silver chloride reference electrode consists of a platinum base coated with silver and silver chloride immersed in a hydrochloric acid solution. Its use is common where mercury contamination cannot be tolerated. It is a stable, reproducible reference electrode with the following overall reaction:

$$AgCl \ + \ e \ = \ Ag \ + \ Cl^- \tag{5.71}$$

The reversible potential is given by an equation of the same form as for the calomel electrode, but because cations are different, the standard reversible potential is not the same.

$$E \ = \ 0.222 \ - \ 0.059 \log a_{Cl^-} \tag{5.72}$$

The temperature coefficient is approximately -0.6 mV/°C.

In applications where chloride contamination is undesirable or where sulfate ions are present, several sulfate-based reference electrodes are suitable for use as reference electrodes. A lead amalgam-lead sulfate is one such system. The mercury-mercurous sulfate electrode is the sulfate analog of the calomel electrode.

$$Hg_2SO_4 + 2e = 2Hg + SO_4^{2-} \qquad (5.73)$$

The reversible potential is given by

$$E = 0.615 - \frac{2.303RT}{2F} \log a_{SO_4^{2-}} \qquad (5.74)$$

When making measurements in basic solution, the mercury-mercuric oxide reference electrode is frequently used. The electrode reaction is

$$HgO + H_2O + 2e = Hg + 2OH^- \qquad (5.75)$$

The reversible electrode potential at 25°C is

$$E = 0.0986 - 0.059 \log a_{OH^-} \qquad (5.76)$$

Potential measurements in nonaqueous media are also possible using reference electrodes. Attempts to relate potentials in different media to each other have led to practical and theoretical difficulties. A major problem is associated with determining the potential at the junction between two different electrolyte solutions. The present convention is to use the SHE in each medium as the primary reference. Although this convention does not allow direct comparison of the same electrode in two different media, it is frequently the case that the same ordering of reversible potentials is maintained. The reversible potentials of several electrode couples in various nonaqueous media are tabulated in Table 5.2.

Use of reference electrodes to measure potential in solution can be illustrated through an example of the deposition of nickel from acid electrolyte. In a typical sulfate bath the primary constituents are $NiSO_4 \cdot 7\,H_2O$ (250 g/L), $NiCl_2 \cdot 6\,H_2O$ (45 g/L), and H_3BO_4 (30 g/L). Typical operating conditions are T = 60°C and pH = 5. An SCE near the working electrode reads 0.67 V. We know that hydrogen will be evolved as a parasitic reaction, and we would like to estimate the current efficiency.

Electrode	H_2O	CH_3OH	C_2H_5OH	NH_3
	25°C	25°C	25°C	-50°C
Cl_2/Cl^-	1.36	1.12	1.05	1.28
Ag/Ag^+	0.80	0.76	0.75	0.83
H_2/H^+	0.00	0.00	0.00	0.00
Tl/Tl^+	-0.34	-0.38	-0.34	–
Na/Na^+	-2.71	-2.73	-2.66	-1.84

Table 5.2: Reversible electrode potentials in different media

Under these conditions the reversible potentials are given by

$$E_{Ni} = -0.23 - 0.015 \ln \left(\frac{1}{[Ni^{2+}]} \right)$$
$$E_{H_2} = -0.059 \, pH$$

Surface overpotentials are given by

$$\eta_{Ni} = 0.06 \log \left(\frac{i}{10^{-5}} \right)$$
$$\eta_{H_2} = 0.1 \log \left(\frac{i}{10^{-4}} \right)$$

Under the specified conditions the reversible potential of the SCE is 0.22 V vs. SHE.

We can calculate the current densities from kinetic expressions for the hydrogen and nickel reactions after we calculate overpotentials from Eqs. 5.66 - 5.68. Under these conditions the concentration of nickel is approximately 1 M. For nickel deposition

$$\begin{aligned} E_{wr,rev} &= E_{w,rev} - E_{r,rev} \\ &= -0.23 - (0.22) \\ &= -0.45 \text{ V} \end{aligned}$$

The overpotential from Eq. 5.68 is

$$\begin{aligned} \eta_{Ni} &= E_{meas} - E_{wr,rev} \\ &= -0.67 - (-0.45) \\ &= -0.22 \text{ V} \end{aligned}$$

The Tafel expression for nickel deposition can be solved explicitly for the current density

$$
\begin{aligned}
i_{Ni} &= 10^{-5} \, 10^{-\eta_{Ni}/0.06} \\
&= 10^{-5} \, 10^{0.22/0.06} \\
&= 0.047 \text{ A/cm}^2
\end{aligned}
$$

An analogous treatment for the hydrogen reaction yields

$$
\begin{aligned}
E_{H_2} &= -0.295 \text{ V} \\
E_{wr,rev} &= -0.052 \text{ V} \\
\eta_{H_2} &= -0.15 \text{ V} \\
i_{H_2} &= 0.003 \text{ A/cm}^2
\end{aligned}
$$

The current efficiency is given by

$$
\begin{aligned}
\epsilon_c &= i_{Ni}/i_{tot} \\
&= 0.047/0.050 \\
&= 0.94
\end{aligned}
$$

The current efficiency is 94% under these conditions.

We can make a number of observations from this example. In practice, the bath is kept at a relatively high pH. Note that a lower pH would increase the reversible hydrogen potential, reduce the hydrogen overpotential, and increase the fraction of hydrogen evolved, provided that all other factors remained the same. At a pH of zero, essentially all of the current would produce hydrogen. At much higher pH values, some of the nickel would precipitate as $Ni(OH)_2$. Higher nickel concentration shifts the reversible potential for nickel deposition to more positive values, which tend to increase current efficiency. Because of the larger Tafel slope for the hydrogen reaction, higher current densities also increase current efficiency.

In this particular case we were able to calculate reversible potentials from Nernst expressions. Often, the formation of complexes in solution significantly alters the activity of individual components. For example, brass can be electrodeposited from a solution of copper and zinc, despite the fact that their standard reversible potentials are separated by 1.1 V. To effect the co-deposition of the two metals, cyanide ions can be added to the solution, and the complexes formed have an activity that causes their reversible potentials to be much closer to each other. In addition, the kinetics of the copper reaction are more sluggish, which also promotes a greater percentage of zinc inclusion in the alloy.

5.6   Mechanistic Studies

A detailed understanding of an electrochemical reaction proceeds from a knowledge of its mechanism. From mechanistic studies a more direct attack on the rate-limiting process is possible. Determining whether adsorption, charge transfer, dissolution, or some other process is retarding the overall rate can be used as a basis for developing a strategy for changing the reaction rate.

A reaction mechanism usually proceeds from a set of postulated elementary reactions. If adsorbed species or film formation are involved, models for these processes must also be included. For thin films a Langumir treatment may be sufficient. For steps involving charge transfer, there is little probability that more than one electron will interact with an ionic, atomic, or molecular species; therefore, only single electron transfers are usually considered in a plausible mechanism. Once a mechanism has been postulated, a test of the results against experimental data is used to determine the plausibility of the model. Because kinetic constants cannot generally be calculated with any accuracy a *priori*, tests of functionality are most appropriate. Quantities such as reaction order, Tafel slope, and variation in exchange current density with potential are commonly used as test variables.

If a proposed mechanism involves a large number of steps, the analysis can be simplified if the rate-limiting steps can be identified. We can simply guess which steps are rate-limiting and determine their effect on the measured quantities. If one set of guesses produces the proper values for the Tafel slope, reaction orders, and other variables, it can be tentatively adopted. Alternatively, we can use insight gained from analogous reaction schemes to establish a more methodical procedure for determining rate-limiting behavior. The remaining reactions are then assumed to be essentially at equilibrium.

If the empirical results are consistent with the postulated mechanism, it is likely that the mechanism is the correct one; however, there are cases where two mechanisms can give the same macroscopic result. A positive test of a mechanism usually relies on the use of analytical techniques to confirm the presence of reaction products or intermediates.

For the case of a single electron transfer, we have developed equations relating current, potential, and species concentrations. Functional relationships follow from these equations. Differentiation of surface overpotential with respect to the logarithm of current density in the Tafel equation (Eq. 5.58) leads to

$$\left(\frac{\partial \eta_s}{\partial \log i}\right)_{T,P,c_i} = \frac{2.303\, RT}{\alpha F} \tag{5.77}$$

This expression is only valid when the reverse reaction can be ignored. Some authors define the Tafel slope in terms of a potential (with respect to a specified reference) rather than an overpotential. In principle, overpotential is a function of the equilibrium potential, which is concentration-dependent. Because concentration at the electrode surface varies with current density, reversible potential also varies. This distinction is usually ignored except in precise work.

Because the values of the anodic and cathodic transfer coefficients are not necessarily equal, there can be two different values for the Tafel slope. These values are determined by substituting α_a or α_c into Eq. 5.77. If the symmetry factor is set equal to 0.5, then the transfer coefficients are equal. In this case the Tafel slope is 118 mV/decade at room temperature.

In ordinary chemical reactions, the reaction order for a reactant is defined in terms of reaction rate. For electrochemical reactions the rate corresponds to a current density, $r = i/nF$.

$$i = nF\left(k_f c_A^a \exp\frac{\alpha_a F}{RT}\eta_s - k_c c_B^b \exp\frac{-\alpha_c F}{RT}\eta_s\right) \tag{5.78}$$

We can define an electrochemical reaction order for a species A as

$$\left(\frac{\partial \log i}{\partial \log c_A}\right)_{T,P,c_j,\phi} = a \tag{5.79}$$

where a is a constant corresponding to the exponent on concentration term. In this case it is the reaction order pertaining to the anodic process. For the cathodic reaction involving the species B, the analogous electrochemical reaction order is

$$\left(\frac{\partial \log i}{\partial \log c_B} \right)_{T,P,c_j,\phi} = b \qquad (5.80)$$

Measurement of electrochemical reaction orders is easiest at potentials where the reverse reaction does not interfere.

Analogous equations apply for the variation in the exchange current density with concentration. For example,

$$\left(\frac{\partial \log i_0}{\partial \log c_i} \right) = p \qquad (5.81)$$

We assume that the electrode potential and concentrations of other species are maintained at constant values. In practice, it is difficult to operate under these conditions because the concentrations at the surface depend on potential.

The above treatment is generally valid for a single-electron transfer. For multiple-electron transfer reactions, a general treatment is not possible. The form of the overall equation may not give constant values for the Tafel slope, reaction order, and other quantities over a range of potential. It may be possible, however, to test a model over a limited potential range, where certain reactions dominate.

As an example of a multistep reaction, consider the anodic dissolution of zinc in alkaline electrolyte. A plot of current density vs. potential appears in Fig. 5.16. This plot is qualitatively different from the one based on the Butler-Volmer equation (Fig. 5.10). It is clear that the Butler-Volmer equation cannot be made to fit this curve. The sharp reduction in current near -1 V vs. SHE as the potential is increased is characteristic of passive behavior.

Near the reversible potential the following mechanism for the anodic reaction was proposed [4]:

$$\text{Zn} + \text{OH}^- = \text{ZnOH} + e \qquad (5.82)$$
$$\text{ZnOH} + 2\text{OH}^- = \text{Zn(OH)}_3^- + e \quad rds \qquad (5.83)$$
$$\text{Zn(OH)}_3^- + \text{OH}^- = \text{Zn(OH)}_4^{2-} \qquad (5.84)$$

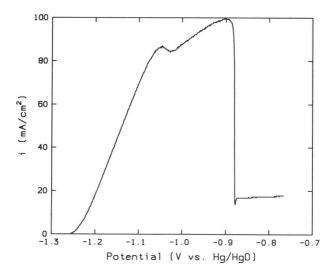

Figure 5.16: Currrent density vs. potential for zinc dissolution on a rotating disk electrode at 25°C, 1200 rpm, 1 N KOH, and 4 mV/s sweep rate.

where *rds* indicates the rate-determining step. In this mechanism we have multiple reaction steps, multiple electron transfers, adsorbed species, and chemical as well as electrochemical steps. As mentioned above, we can simply guess that a particular step is rate-determining and carry out the analysis. Any result that contradicts observed results can be discarded. To carry out the analysis in this case, we treat the second step as the rate-determining process.

To simplify the nomenclature, we designate each species with a single letter: hydroxide $H = OH^-$, zincate $Z = Zn(OH)_3^-$, and tetrahydroxo zincate $T = Zn(OH)_4^{2-}$. The fraction of the surface covered with ZnOH is θ, and the uncovered fraction, consisting of zinc, is $(1 - \theta)$. The reaction numbers 1, 2, and 3 correspond to reactions 5.82, 5.83, and 5.84, respectively. The subscript a is for the anodic process, and the subscript c refers to the cathodic process. Using Eqs. 5.39 and 5.40 as models for elementary electrochemical processes, we can write the rate expressions for the reactions

$$r_1 = k_{a1}c_H(1-\theta)\exp\left[\frac{(1-\beta_1)F\phi}{RT}\right] - k_{c1}\theta\exp\left(\frac{-\beta_1 F\phi}{RT}\right) \quad (5.85)$$

$$r_2 = k_{a2}c_H^2\theta\exp\left[\frac{(1-\beta_2)F\phi}{RT}\right] - k_{c2}(1-\theta)c_Z\exp\left(\frac{-\beta_2 F\phi}{RT}\right) \quad (5.86)$$

$$r_3 = k_{a3}c_Z c_H - k_{c3}c_T \quad (5.87)$$

Note that the term $(1-\theta)$ enters into the cathodic portion of Eq. 5.86 because the $Zn(OH)_3^-$ from solution can only form ZnOH on an uncovered portion of the electrode surface; our assumption of monolayer coverage does not permit additional formation of ZnOH on a portion of the surface already occupied by that species.

   This reaction scheme represents three reactions in series; therefore, at steady state the net, forward reactions proceed at the same rate. One electron is transferred in each of the electrochemical reactions. Because a total of two electrons are transferred, each electrochemical reaction accounts for half of the total current, or

$$i = 2i_1 = 2i_2 \quad (5.88)$$

Reaction 5.82 is assumed to be essentially at equilibrium, and we equate the forward and backward rates in Eq. 5.85 to calculate θ.

$$\theta = \frac{k_{a1}c_H}{[k_{a1}c_H + k_{c1}\exp(-F\phi/RT)]} \quad (5.89)$$

Substitution of Eq. 5.89 into Eq. 5.86 yields

$$\frac{i}{2F} = \frac{k_{a2}k_{a1}c_H^3\exp[(2-\beta_2)F\phi/RT]}{(J+1)k_{c1}}$$
$$- \frac{k_{c2}k_{c3}c_T\exp(-\beta_2 F\phi/RT)}{(J+1)k_{a3}c_H} \quad (5.90)$$

where

$$J = \frac{k_{a1}c_H}{k_{c1}\exp(-F\phi/RT)}$$

At equilibrium we set the current density to zero in Eq. 5.90 to calculate the equilibrium potential

$$\phi^o = \frac{RT}{2F}\ln\frac{k_{c1}k_{c2}k_{c3}c_T}{k_{a1}k_{a2}k_{a3}c_H^4} \quad (5.91)$$

This equation can be expressed more compactly in terms of equilibrium constants, which are ratios of forward to backward rate constants.

$$\phi^o = \frac{RT}{2F} \ln \frac{c_T}{K_1 K_2 K_3 c_H^4} \tag{5.92}$$

The exchange current density is proportional to the rate in the anodic or cathodic direction at the equilibrium potential. To evaluate this quantity, we substitute ϕ^o from Eq. 5.91 into one of the terms in Eq. 5.90. Because we want to calculate the rate in only one direction, we drop the first term on the right-hand side of Eq. 5.90

$$i_o = \frac{2F k_{c2} \left(K_1 K_2 \right)^{\beta_2/2} K_3^{(\beta_2-2)/2} c_H^{2\beta_2-1} c_T^{(2-\beta_2)/2}}{J^o + 1} \tag{5.93}$$

where J^o is the variable J evaluated at the equilibrium potential ϕ^o. The overpotential represents a departure from the equilibrium potential.

$$\eta_s = \phi - \phi^o \tag{5.94}$$

For this reaction mechanism we can cast the current density-overpotential expression in a form similar to the Butler-Volmer equation. Substitution of Eqs. 5.91 and 5.94 into Eq. 5.90 yields

$$i = i_0^* \left[\exp \left(\frac{\alpha_a F \eta_s}{RT} \right) - \exp \left(\frac{-\alpha_c F \eta_s}{RT} \right) \right] \tag{5.95}$$

where

$$i_0^* = i_0 \frac{J^o + 1}{J + 1}$$
$$\alpha_a = 2 - \beta_2$$
$$\alpha_c = \beta_2$$

This rate equation differs from the standard Butler-Volmer equation in the potential dependence of the pre-exponential term. As we noted earlier, a Butler-Volmer type equation cannot simulate current-overpotential over a large potential range. This expression only agrees with experimental results within 100 mV of the rest potential; however, the predicted reaction orders and Tafel slopes agree with observed values.

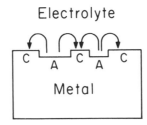

Figure 5.17: Schematic of a corrosion cell. Regions labeled A represent anodic areas where metal is dissolving. Regions labeled C represent cathodic areas. Vectors represent lines of current flow.

5.7   The Kinetics of Corrosion Processes

In corroding systems, material is degraded through anodic oxidation. Often cathodic and anodic sites are on the same—usually metallic—surface. The individual sites are small areas that may change position with time. The net effect is that of a short circuited battery as shown in Fig. 5.17. Current flows in the electrolyte adjacent to the metal and electrons flow through the metal. The reactions occurring on the cathodic and anodic portions of corroding surfaces are invariably different, and the lines of current flow do not correspond to simple metal transport from anodic to cathodic sites. For example, on a corroding iron sample, iron would dissolve from the anodic sites and oxygen reduction might occur on the cathodic sites.

Because of charge conservation, we know that the total anodic current must equal the total cathodic current, or

$$I_a = I_c \tag{5.96}$$

For the corrosion of a single metal in an electrolyte, we have the anodic and cathodic sites distributed on the surface, but we cannot easily determine the surface area associated with each site. A common assumption is that cathodic and anodic sites occupy approximately the same area when metal is freely corroding. In aqueous electrolyte a cathodic reaction is often the reduction of molecular oxygen or hydrogen ions.

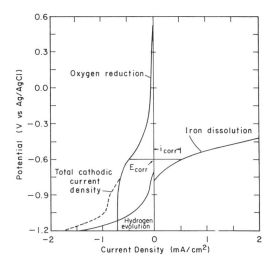

Figure 5.18: Polarization curve for iron in aerated seawater. Corrosion potential occurs at the location where anodic and cathodic currents balance.

5.7.1 The Evans Diagram

If we plot polarization curves for the anodic and cathodic curves, we can estimate the current and potential of a freely corroding metal. Such a plot for iron in flowing seawater is shown in Fig. 5.18. On the potential scale that we have chosen, the reversible potential of iron is roughly -0.7 V and the reversible potential for the reduction of oxygen is approximately 0.4 V. For any electrochemical process involving Faradaic reactions, charge conservation (Eq. 5.96) must be maintained. If we assume that the anodic sites are equal in area to the cathodic sites, then we can see from Fig. 5.18 that the potential of the metal must be at the point where the anodic and cathodic currents are equal. For this system the point is labeled E_{corr}, the corrosion potential, and the corresponding current is i_{corr}.

Because anodic and cathodic processes are occurring simultaneously on a corroding surface, the separation of the individual processes is not readily performed. At the corrosion potential, all of the current is being transferred on the same electrode surface, and the current in the external circuit is zero. There are several standard techniques for estimating the rates of individual processes near the corrosion potential. The weight loss of a corroding sample can be monitored as a function of time. Through Faraday's law the weight loss can be directly related to the charge passed and to the reaction rate. A second technique is to assume that the processes far from the corrosion potential follow the same rate law near the corrosion potential. Often a Tafel relation is assumed far from the E_{corr}; hence, this technique is referred to as Tafel extrapolation. Another method is to measure the reaction rate of one constituent after removing the other reactive component. For example, we could deaerate the solution and measure the anodic current due to the dissolution of the metal.

The measurement of E_{corr} is straightforward. In practice, we attach the lead of a reference electrode to the corroding material and place the electrode tip in solution near the material. If we are measuring the corrosion potential of a single material, then the current is concentrated near the electrode surface, and the ohmic drop at points removed from the metal surface is minimal. We can, for example, measure the corrosion potential of a buried structure (pipeline or storage tank) from remote reference electrode readings near the soil surface.

The kinetics of corroding systems are often presented in simplified form on plots known as Evans diagrams (see Fig. 5.19). The main assumption is that the behavior over the range of interest follows a Tafel relation, i.e., a plot of the potential vs. logarithm of current density is linear. By plotting the absolute value of the current density, we can determine the corrosion potential and current density directly from the diagram.

More complex systems are sometimes depicted on an Evans diagram, but such diagrams are of less value when the current density is nonuniform. Consider the case where zinc and iron are in contact in an electrolyte (Fig. 5.20). Qualitatively, we have current flowing from the zinc to the iron. This type of arrangement causes the zinc to dissolve, but it lowers the potential adjacent to the metal to a value below its corrosion potential. If the potential is lowered below the reversible potential of the iron, it will no longer corrode; this phenomenon is the basis of cathodic protection.

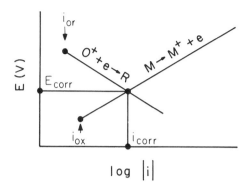

Figure 5.19: Schematic Evans diagram for a corroding system. Exchange current densities for oxidation and reduction reactions are i_{0x} and i_{0r}, respectively.

The corresponding Evans diagram is shown in Fig. 5.21. The intersection of the zinc and hydrogen curves show the behavior expected along the line of intimate contact between the two metals. For the iron alone in electrolyte, the corrosion current is i'_{corr} and the iron being oxidized. When the zinc is in contact, the potential is lower, and the iron ions, if present, are reduced. The corrosion current is now i_{corr}, and it is only the zinc that is corroding. If iron ions were present, they would be reduced and the iron reduction current would add to the hydrogen reduction current to give the total cathodic cur-

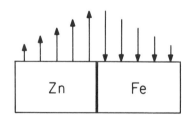

Figure 5.20: Galvanically-coupled iron and zinc samples. Vectors pointing outward represent anodic currents; those pointing toward the metal represent cathodic currents.

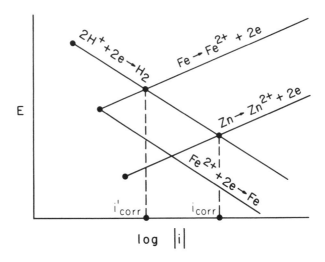

Figure 5.21: Evans diagram for the iron-zinc system. The corrosion current changes from i'_{corr} to i_{corr} when the zinc is attached.

rent. Because the current is displayed on a logarithmic scale, the iron reduction current would be several orders of magnitude lower than the hydrogen reduction current and could be ignored in calculations.

Farther away from the junction, the current must flow through longer electrolyte paths with correspondingly larger ohmic drops. Schematically, the current distribution is as shown in Fig. 5.20. The current distribution depends on a number of factors including electrolyte conductivity, Tafel slope, and system geometry. Of these factors only the Tafel slope is taken into account on the Evans diagram. Away from the junction of the two metals, we have no way of determining the potential adjacent to the metal from the Evans diagram. Multiple reactions can also be displayed on the Evans diagram. When iron is coupled to zinc, the potential adjacent to the iron may be sufficiently low to reduce both oxygen and hydrogen ions. While this behavior can be easily displayed on an Evans diagram, the information is most useful when the current density is uniform. Approximate equations for determining the potential distribution are available for simple geometries. More generally, a numerical technique would be required to simulate the current and potential distributions.

5.7.2 Simplified Kinetic Expressions

For several special cases corrosion kinetics can be reduced to simplified forms [5]. If we consider the case of two reactions following Butler-Volmer kinetics, then the current densities of the oxidation and reduction processes are equal at the corrosion potential.

$$
\begin{aligned}
i_{0x} &\left[\exp \frac{\alpha_{ax} F}{RT} (E_{corr} - E_x) - \exp \frac{-\alpha_{cx} F}{RT} (E_{corr} - E_x) \right] \\
&= i_{0r} \left[\exp \frac{\alpha_{ar} F}{RT} (E_{corr} - E_r) - \exp \frac{-\alpha_{cr} F}{RT} (E_{corr} - E_r) \right] \quad (5.97)
\end{aligned}
$$

The subscript x refers to the oxidation process, the subscript r refers to the reduction process; E_x and E_r refer to the reversible potentials of the oxidation and reduction potentials, respectively. If all of the transfer coefficients are taken to be equal to one-half, then we can calculate a corrosion potential directly.

$$
E_{corr} = \frac{RT}{F} \ln \left[\frac{i_{0x} \exp(F/2RT) E_x + i_{0r} \exp(F/2RT) E_r}{i_{0x} \exp(-F/2RT) E_x + i_{0c} \exp(-F/2RT) E_r} \right]
$$
$$(5.98)$$

Another simplified equation results if we have a system where the reversible potentials of the oxidation and reduction reactions are far removed from the corrosion potential. This result occurs when the Tafel approximation is applied to both reactions. In that case we can neglect the back reaction for each process, and Eq. 5.97 simplifies to

$$
i_{corr} = i_{0x} \exp \frac{\alpha_{ax} F}{RT} (E_{corr} - E_x) \quad (5.99)
$$

$$
= i_{0r} \exp \frac{-\alpha_{cr} F}{RT} (E_{corr} - E_r) \quad (5.100)
$$

Here we are only interested in the magnitude of the corrosion current density, and we treat each component as a positive quantity. Equating the right-hand sides of the above two equations, we obtain the following expression for the corrosion potential:

$$
E_{corr} = \frac{(RT/F)\ln(i_{0r}/i_{0x}) + \alpha_{cr} E_r + \alpha_{ax} E_x}{\alpha_{ax} + \alpha_{cr}} \quad (5.101)
$$

If the transfer coefficients are assumed to be equal to one-half, the above equation reduces to

$$E_{corr} = \frac{RT}{F} \ln \frac{i_{0r}}{i_{0x}} + \frac{E_x + E_r}{2} \qquad (5.102)$$

When the exchange current densities for the reactions are equal, the logarithmic term becomes zero, and the corrosion potential is simply the average of the reversible potentials of the two reactions.

By substituting Eq. 5.101 into Eq. 5.99, we obtain an expression for the corrosion current density in terms of the kinetic parameters and reversible potentials.

$$i_{corr} = i_{0x}^{\alpha_{cr}/(\alpha_{ax}+\alpha_{cr})} i_{0r}^{\alpha_{ax}/(\alpha_{ax}+\alpha_{cr})}$$
$$\exp\left[\frac{F}{RT}\frac{\alpha_{ax}\alpha_{cr}}{\alpha_{ax}+\alpha_{cr}}(E_r - E_x)\right] \qquad (5.103)$$

If the transfer coefficients are taken to be equal to one-half, then the following simplified expression results:

$$i_{corr} = (i_{0x}i_{0r})^{1/2} \exp\left[\frac{F}{4RT}(E_r - E_x)\right] \qquad (5.104)$$

The kinetic expressions derived above are valid at the corrosion potential and corresponding current density. For certain applications, such as cathodic protection, we are interested in operating at a potential ϕ different from the corrosion potential, and we want to calculate the corresponding current density. The overpotential is the deviation from the reversible potential, or

$$\eta_s = \phi - E_x \qquad (5.105)$$

An analogous equation applies to the deviation from E_r. To obtain a current-potential expression in the desired form, we can add and subtract E_{corr} from the right-hand side of the above equation. If we assume that ϕ is far from the reversible potential of either reaction, then we can use the Tafel approximation for the oxidation and reduction reactions.

$$i = i_{0x} \exp\left[\frac{\alpha_{ax}F}{RT}(\phi - E_{corr} + E_{corr} - E_x)\right]$$
$$-i_{0r} \exp\left[\frac{-\alpha_{cr}F}{RT}(\phi - E_{corr} + E_{corr} - E_r)\right] \qquad (5.106)$$

Rearranging gives

$$i = i_{0x} \exp\left[\frac{\alpha_{ax}F}{RT}(E_{corr} - E_x)\right] \exp\left[\frac{\alpha_{ax}F}{RT}(\phi - E_{corr})\right]$$
$$-i_{0r} \exp\left[\frac{-\alpha_{cr}F}{RT}(E_{corr} - E_r)\right] \exp\left[\frac{-\alpha_{cr}F}{RT}(\phi - E_{corr})\right] (5.107)$$

By substituting Eqs. 5.99 and 5.100 into Eq. 5.107, we can express the current density in terms of the corrosion current density.

$$i = i_{corr}\left[\exp\frac{\alpha_{ax}F}{RT}(\phi - E_{corr}) - \exp\frac{-\alpha_{cr}F}{RT}(\phi - E_{corr})\right] \quad (5.108)$$

Although this equation is similar in form to the Butler-Volmer equation, it should not be regarded as giving the same relationship, i.e., the current attributable to a departure from an equilibrium potential.

Small potential excursions around the corrosion potential are sometimes applied in an attempt to extract kinetic parameters. If $\phi - E_{corr}$ is small, we can expand the exponential terms from Eq. 5.107 in a Maclaurin series by

$$i = i_{corr}\left[\left(\frac{\alpha_{ax}F}{RT} + \frac{\alpha_{cr}F}{RT}\right)(\phi - E_{corr})\right] \quad (5.109)$$

Terms involving the transfer coefficients are inversely related to the Tafel slopes, or

$$B_x = 2.303\frac{RT}{\alpha_{ax}F} \quad (5.110)$$

$$B_r = 2.303\frac{RT}{\alpha_{cr}F} \quad (5.111)$$

Substitution of the above definitions into Eq. 5.109 yields

$$i = 2.303\, i_{corr}\left(\frac{B_x + B_r}{B_x B_r}\right)(\phi - E_{corr}) \quad (5.112)$$

This expression is referred to as the Stern-Geary relation [6].

The calculation of system kinetics in corroding systems is complicated by a number of factors. We have not yet addressed mass transport limitations. For example, when oxygen reduction is the

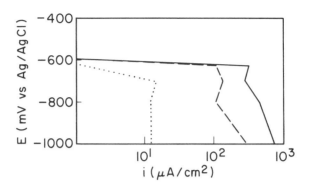

Figure 5.22: Time-dependence of polarization curves for steel in flowing seawater. Solid line represents data at 5 min.; dashed line 1 day; dotted line 120 days.

cathodic reaction, mass transport often limits the overall rate. The solubility of oxygen in water is on the order of 10 ppm, and the mass transport-limited current density is less than 0.01 mA/cm² unless agitation is vigorous. Because of changes in surface character or the formation of surface deposits, the polarization curve and corrosion potential can change significantly over time [7], as shown in Fig. 5.22. In the above treatment we have only considered uniform corrosion. A further complication results from non-homogeneous (localized) attack. For example, the formation of a pit can cause the failure of an entire structure, even though the amount of metal dissolved is very small. Localized corrosion is a separate subject; a primer on the subject has been presented by Fontana [8].

5.8   Problems

1. When the dimensionless potential $\psi \ll 1$ in Eq. 5.17, the exponentials in the hyperbolic sine term can be expanded in a Maclaurin series, and higher order terms neglected.
a) Carry out this expansion, and solve the resulting linear differential equation subject to the same boundary conditions (Eqs. 5.20 and 5.21).

b) Plot the potential vs. dimensionless distance χ for the approximate and more general solution (Eq. 5.27) when $\psi_0 = 1$.

2. The kinetics of an electrochemical reaction are described by the Butler-Volmer equation with the parameters $\alpha_a = \alpha_c = 0.5$ and $i_0 = 10$ mA/cm^2.

a) Determine the error in using the Tafel approximation for overpotentials of 1 mV and 100 mV.

b) Repeat Part a for the linear approximation.

3. Overpotential measurements at 25°C for copper dissolution in well-stirred electrolyte yielded the following results:

i (mA/cm^2)	1.2	2.4	4.8	9.7	20	40	60	200	2000
η_s (mV)	1.5	3.0	6.0	9.0	18	30	36	60	104

Determine i_0 and α_a.

4. An SCE is placed near an inert electrode where hydrogen is evolving from pH 5 electrolyte. At 25°C the following potentials are measured with a voltmeter connected between the reference electrode and cathode:

i (mA/cm^2)	10	1	0.3	0.1
E_{meas} (V)	-0.91	-0.79	-0.71	-0.66

a) Calculate the cathodic overpotential for each measurement.

b) Estimate i_0 and α_c from these measurements.

5. Two large plane parallel copper electrodes are immersed in a 1 M solution of $CuSO_4$ at 25°C. The gap between the electrodes is 1 cm. Estimate the potential difference between the anode and cathode (applied potential) when the average current density is 1 mA/cm^2.

6. A reaction follows Butler-Volmer kinetics with $\alpha_a = \alpha_c = 0.5$ and $i_0 = 1$ ma/cm^2.

a) Determine the increase in reaction rate when the overpotential is increased from 0.1 V to 1.1 V.

b) For a chemical reaction following Arrhenius behavior, determine the temperature required to achieve the same increase. Assume that the rate is initially measured at 25°C and that the activation energy is 100 kJ.

7. For the following reaction:

$$I_3^- + 2e = 3I^-$$

a three-step mechanism has been proposed:

$$I_3^- = I_2 + I^-$$
$$I_2 = 2I$$
$$2(I + e = I^-)$$

The observed anodic reaction order is 1 for I^-. The observed cathodic reaction orders are - 1/2 for I^- and 1/2 for I_3^-.

a) Why is the last reaction written as occurring twice with a single electron transfer?

b) Determine which of the three steps is rate-limiting. Treat the remaining two steps as equilibrium reactions and calculate the reaction orders. For a plausible mechanism the calculated reaction orders should agree with the observed reaction orders.

8. Zinc is being deposited from acid electrolyte. The concentration of $ZnCl_2$ is 1 M and the pH is 3. At 25°C both zinc deposition and hydrogen evolution are possible. Assume that the reactions follow Tafel behavior in the region of interest. For the zinc reaction $\alpha_c = 0.5$ and $i_0 = 10^{-1}$ A/cm². For hydrogen $\alpha_c = 0.5$ and $i_0 = 10^{-9}$ A/cm². A voltmeter connected to an SCE reference electrode reads -1.2 V.

a) Estimate the current efficiency for zinc deposition.

b) In qualitative terms explain the effect of raising the pH on the current efficiency.

9. The potential of a metal oxidation reaction is given by

$$E = -0.2 + 0.06 \log \frac{i}{10^{-6}}$$

The potential of a reduction reaction in the same solution is

$$E = 0.1 - 0.1 \log \frac{i}{10^{-7}}$$

If these are the only two reactions possible on the metal surface, estimate the corrosion potential and current density.

Bibliography

[1] D. C. Grahame, *J. Am. Chem. Soc.*, **76**, 4819 (1954).

[2] L. I. Antropov, *Theoretical Electrochemistry* (Moscow: Mir Publishers), p. 418.

[3] D. J. Ives and G. J. Janz, eds. *Reference Electrodes* (New York: Academic Press, 1961).

[4] Yu-Chi Chang and G. A. Prentice, *J. Electrochem. Soc.*, **131**, 1465 (1984).

[5] W. H. Smyrl, "Electrochemistry and Corrosion on Metal Surfaces," *Comprehensive Treatise of Electrochemistry*, **4**, 97 (1981).

[6] M. Stern and A. L. Geary, *J. Electrochem. Soc.*, **104**, 56 (1957).

[7] J. R. Scully, H. P. Hack, and D. G. Tipton, *Corrosion*, **42**, 462 (1986).

[8] M. G. Fontana, *Corrosion Engineering* (New York: McGraw- Hill Book Co., 1975), Chap. 3.

Chapter 6

Ionic Mass Transport

Charge transfer at electrode surfaces is accompanied by the transport of ions in the electrolyte. When uncharged species are involved, their transport is governed by the ordinary laws of convection and diffusion. For charged species, electric field effects must also be considered. In a rigorous treatment of transport processes, we would calculate species fluxes from the gradient of the electrochemical potential. In practice, fluxes are calculated from readily measurable quantities, and a dilute solution treatment is often used for more concentrated solutions.

In engineering applications electrochemical processes are often limited by the rate of reactant transport to the electrode surface. The mass transport-limited rate can usually be increased through the application of several techniques: increasing the agitation of the fluid, increasing the concentration of the reactant, or increasing the temperature of the solution.

For a few well-behaved systems, we can rigorously calculate the mass transport rate from analytical solutions to the relevant equations. By well-behaved, we mean laminar flow in cells displaying a high degree of symmetry. In oddly shaped cells or in cells with turbulent flow, there is little chance of obtaining an analytical result, and we must resort to numerical techniques.

Because ion fluxes are directly related to current density, a solution to the current distribution problem results directly from a solution to the mass transport problem. In modeling electrochemical systems, we want to determine which effects can be justifiably neglected. In

151

some cases we can neglect concentration gradients because the system
is governed by electric field effects; in such cases we are then solving
problems from potential theory.

6.1   Fundamental Relationships

A general description of an electrochemical system takes into account
the following effects: species fluxes, material conservation, current
flow, electroneutrality, electrode kinetics, and hydrodynamics. It is
usually possible to identify the pertinent equations, but obtaining a
solution may not be feasible unless suitable approximations are made.
Engineering judgment is required in simplifying equations. Expressions such as the conservation equations reflect fundamental physical
concepts and cannot be ignored or altered. Other expressions reflect
observed behavior and are subject to simplification. The flux and
kinetic expressions are in this latter category.

One common description of an electrochemical system is derived
from assuming that the concentrations of electroactive species are
extremely small. In this treatment only interactions between the solute and solvent are considered, and solute-solute interactions are ignored. Under these conditions an expression for describing the flux of
a species is

$$\mathbf{N}_i \; = \; -z_i u_i F c_i \nabla \phi \; - \; D_i \nabla c_i \; + \; c_i \mathbf{v} \qquad (6.1)$$

where $\mathbf{N}_i$ is the flux, z_i is the charge on the ion, u_i is the mobility, c_i is
the species concentration, $\nabla \phi$ is the potential gradient, D_i is the diffusivity, and $\mathbf{v}$ is the bulk fluid velocity. Flux, potential gradient, and
fluid velocity are vector quantities, indicated by boldface type. The
terms on the right side of the equation represent fluxes resulting from
migration, diffusion, and convection, respectively. In a strict sense
this equation is only valid for infinitely dilute solutions; however, this
formulation is commonly extended to more concentrated solutions. Its
use is particularly convenient because transport properties are usually
tabulated in terms of the parameters found in Eq. 6.1 or parameters
directly related to them. Current arises from the motion of all charged

species

$$\mathbf{i} = F \sum_i z_i \mathbf{N}_i \tag{6.2}$$

The material balance in the bulk electrolyte can be expressed as

$$\frac{\partial c_i}{\partial t} = -\nabla \cdot \mathbf{N}_i + R_i \tag{6.3}$$

where R_i represents a chemical reaction occurring in solution. Because the electrical forces between charged species are so large, significant charge separation cannot occur. Deviations from electroneutrality are only observed in the double-layer region, which can be modeled separately. In the bulk electrolyte the electroneutrality assumption is valid.

$$\sum_i z_i c_i = 0 \tag{6.4}$$

The four preceding equations form a basis for modeling mass transport in electrolytic solutions. To carry out a simulation, we need a description of the bulk velocity, a variable appearing in the flux equation. This description requires a separate solution obtained from the equations of fluid mechanics. The fluid mechanics problem can be avoided by making certain assumptions. For example, if we neglect concentration gradients, then the transport of reactants to the electrode surface cannot be rate-limiting. This condition might be assumed in a well-stirred, low viscosity electrolyte system, where it is approachable.

Solutions to these four equations depend on cell geometry, electrode kinetics, and hydrodynamic conditions. With simplifying assumptions the equations reduce to familiar forms, valid for large classes of problems. If concentration variations and homogeneous reaction can be ignored ($R_i = 0$), the equations reduce to a particularly simple form. Multiplication of Eq. 6.3 by $z_i F$ and addition over all species yield

$$\frac{\partial}{\partial t} F \sum_i z_i c_i = -\nabla \cdot F \sum_i z_i \mathbf{N}_i \tag{6.5}$$

Substitution of Eqs. 6.2 and 6.4 into Eq. 6.5 gives

$$\nabla \cdot \mathbf{i} = 0 \tag{6.6}$$

An expression for current density in terms of the species fluxes is obtained by substituting Eq. 6.1 into Eq. 6.2.

$$\mathbf{i} = -F^2\nabla\phi\sum_i z_i^2 u_i c_i - F\sum_i z_i D_i \nabla c_i + F\mathbf{v}\sum_i z_i c_i \qquad (6.7)$$

The last term on the right side is zero according to the electroneutrality assumption (Eq. 6.4), meaning that ions are transported in electrically neutral combinations by the fluid flow. As we intuitively expect, a flowing fluid containing ions generates no net current. For example, no current can be extracted from an aqueous solution of sodium chloride flowing in a pipe. If concentration gradients can be neglected, then the second term is also zero, and Eq. 6.7 reduces to

$$\mathbf{i} = -\kappa\nabla\phi \qquad (6.8)$$

where

$$\kappa = F^2 \sum_i z_i^2 u_i c_i$$

Eq. 6.8 is an expression of Ohm's law: current density is proportional to the gradient of the potential. Combining Eqs. 6.6 and 6.8 yields

$$\nabla\cdot(-\kappa\nabla\phi) = 0 \qquad (6.9)$$

Because we assume that temperature and concentration are uniform, conductivity is constant, and Eq. 6.9 reduces to Laplace's equation, which is

$$\nabla^2\phi = 0 \qquad (6.10)$$

Another special case of interest is that of a solution of a single salt, the binary electrolyte. To maintain electroneutrality, we must have

$$z_+\nu_+ + z_-\nu_- = 0 \qquad (6.11)$$

where ν_i is the number of ions produced from the dissociation of a single molecule of the salt. The concentration is given by

$$c = \frac{c_+}{\nu_+} = \frac{c_-}{\nu_-} \qquad (6.12)$$

Substitution of Eq. 6.1 into Eq. 6.3 with $R_i = 0$ for each species yields

$$\frac{\partial c}{\partial t} + v \cdot \nabla c = z_+ u_+ F \nabla \cdot (c \nabla \phi) + D_+ \nabla^2 c \qquad (6.13)$$

$$\frac{\partial c}{\partial t} + v \cdot \nabla c = z_- u_- F \nabla \cdot (c \nabla \phi) + D_- \nabla^2 c \qquad (6.14)$$

The assumption that the fluid is incompressible implies that the term $\nabla \cdot v = 0$, and factors involving this quantity do not appear in Eqs. 6.13 and 6.14. Subtracting Eq. 6.14 from Eq. 6.13 gives

$$(z_+ u_+ - z_- u_-) F \nabla \cdot (c \nabla \phi) + (D_+ - D_-) \nabla^2 c = 0 \qquad (6.15)$$

Combining Eqs. 6.15 and 6.13 yields

$$\frac{\partial c}{\partial t} + v \cdot \nabla c = D \nabla^2 c \qquad (6.16)$$

where

$$D = \frac{z_+ u_+ D_- - z_- u_- D_+}{z_+ u_+ - z_- u_-} \qquad (6.17)$$

This is the equation of convective diffusion, which has several useful features. Because potential was eliminated, we could calculate the concentration distribution in a binary electrolyte without calculating the potential distribution; however, the potential distribution is obtainable from Eq. 6.15. The convective diffusion equation has analogs in heat and mass transport, and many solutions are available.

For binary electrolyte the governing differential equation was simplified because we eliminated the potential. A similar situation arises when we have an excess of supporting electrolyte. Substitution of Eq. 6.1 into Eq. 6.3 gives

$$\frac{\partial c_i}{\partial t} + \mathbf{v} \cdot \nabla c_i = z_i F \nabla \cdot (u_i c_i \nabla \phi) + \nabla \cdot (D_i \nabla c_i) \qquad (6.18)$$

To calculate the concentration distribution, we would need to solve a set of equations in the form of Eq. 6.18 for each ionic species. To perform these calculations rigorously in the general case is a complex numerical task. In an excess of supporting electrolyte, the coefficient of the potential gradient term is large; consequently, the potential gradient must be relatively small and is assumed to be negligible. Eq. 6.18 reduces to

$$\frac{\partial c_i}{\partial t} + \mathbf{v} \cdot \boldsymbol{\nabla} c_i = D_i \boldsymbol{\nabla}^2 c_i \qquad (6.19)$$

This equation is of the form of Eq. 6.18 but is more complicated because the supporting electrolyte case requires the simultaneous solution of a set of equations, rather than the solution of a single equation.

6.2   Mass Transport Boundary Layer

For systems where concentration gradients cannot be neglected, rigorous solutions of the partial differential equations governing fluid motion and ion transport can be difficult to obtain; consequently, a number of approximate methods have been developed for use in making engineering estimates. One simplification is to treat the mass transport boundary layer near an electrode surface as a region where the concentration gradient is linear. This method is equivalent to assuming that only diffusion takes place near the electrode and that diffusivity is invariant in this region. The concentration gradient in this fictitious boundary layer is taken to be the gradient corresponding to the flux at the electrode surface (Fig. 6.1). This model of the boundary layer is known as the Nernst diffusion layer. The thickness of this layer is a convenient measure of the resistance to mass transport. For a specified cell configuration the thickness of the boundary layer is a function of system hydrodynamics; a thinner layer corresponds to greater fluid agitation and facilitated mass transport.

As an application of the Nernst diffusion layer treatment, we consider the deposition of a metal ion from binary electrolyte. The flux of the metal cation at a plane electrode surface is due to potential and concentration gradients. Because fluid velocity is zero at the surface, the convective term in Eq. 6.1 is zero, and the flux expression for the cation reduces to

$$N_+ = -z_+ u_+ F c_+ \frac{\partial \phi}{\partial x} - D_+ \nu_+ \frac{\partial c}{\partial x} \qquad (6.20)$$

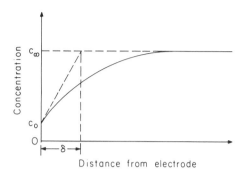

Figure 6.1: Nernst diffusion layer model. The solid line represents the actual concentration profile, and the dashed line from c_0 is the extrapolation of the initial slope.

where x is the distance from the electrode surface. The current is due to the flux of all ions, but at the electrode surface only the cations contribute to the current and Eq. 6.2 becomes

$$N_+ = \frac{i}{z_+ F} \tag{6.21}$$

Anions do not participate in the electrode reaction and their flux at the electrode surface is zero.

$$N_- = 0$$
$$= -z_- u_- F \nu_- c \frac{\partial \phi}{\partial x} - D_- \nu_- \frac{\partial c}{\partial x} \tag{6.22}$$

Combining Eqs. 6.20, 6.21, and 6.22 and eliminating the potential gradient, we obtain

$$\frac{i}{z_+ \nu_+ F} = -\frac{z_- u_- D_+ - z_+ u_+ D_-}{z_- u_-} \frac{\partial c}{\partial x} \tag{6.23}$$

For binary electrolyte the transference number is defined as

$$t_+ = \frac{z_+^2 u_+ c_+}{z_+^2 u_+ c_+ + z_-^2 u_- c_-} \tag{6.24}$$
$$= \frac{z_+ u_+}{z_+ u_+ - z_- u_-} \tag{6.25}$$

The diffusion coefficient is defined as

$$D = \frac{z_+ u_+ D_- - z_- u_- D_+}{z_+ u_+ - z_- u_-} \tag{6.26}$$

Substituting these expressions into Eq. 6.23 yields

$$\frac{i}{z_+ \nu_+ F} = -\frac{D}{1 - t_+} \frac{\partial c}{\partial x} \tag{6.27}$$

For metal deposition the quantity $z_+ \nu_+$ is equal to the number of electrons participating in the reaction n. Eq. 6.28 can now be expressed as

$$i = -\frac{nFD}{1 - t_+} \frac{\partial c}{\partial x} \tag{6.28}$$

Although both the transference number and diffusivity vary with concentration, in this treatment we have assumed that they are constants.

To reduce ohmic losses in a metal deposition process, a supporting electrolyte is often added to the solution. For example, in a copper deposition process, sulfuric acid increases conductivity but does not participate in the reaction. Because hydrogen ions are more mobile and because their concentration can be made much higher than the copper ion concentration, the transference number for copper ions is much smaller than in binary electrolyte. As an approximation we can assume that the transference number for the copper ion is zero, and Eq. 6.28 becomes

$$i = -nFD\frac{\partial c}{\partial x} \tag{6.29}$$

If we use the Nernst diffusion layer model, the concentration gradients are linear, and Eq. 6.29 can be expressed as

$$i = \frac{-nFD(c_\infty - c_0)}{\delta} \tag{6.30}$$

where c_∞ is the bulk copper ion concentration, c_0 is the surface concentration, and δ is the Nernst diffusion layer thickness.

Increasing current density results in a depletion of copper ions immediately adjacent to the electrode surface. When current density increases, the concentration gradient increases in accordance with

Eq. 6.30. If the solution is well-stirred, the deposition process has a negligible effect on the hydrodynamics, and δ remains constant. When the current is increased to a point where the concentration at the surface falls to zero, a further increase in current does not result in an increased deposition rate; rather another reaction commences. The current density corresponding to zero surface concentration is called the limiting current density, i_l.

$$i_l = \frac{-nFDc_\infty}{\delta} \qquad (6.31)$$

If there are several reactive species, there is a limiting current density for each one.

In initially quiescent aqueous electrolyte δ is typically of the order of 0.05 cm; in stirred solution it is typically of the order of 0.001 cm. As discussed below, initially quiescent solutions are driven by density differences, and the resulting flows reduce the boundary layer thickness. In most aqueous systems, density gradients would result from ion depletion in the boundary layer; however, in a truly immobile electrolyte a boundary layer on the order of half the distance between the electrodes would be established at steady state.

For common metal cations near room temperature, the diffusivity is approximately 8×10^{-6}. For unstirred solutions Eq. 6.31 becomes

$$i_l = 0.015\,nc \qquad (6.32)$$

where c is expressed in mol/L. In well-stirred solutions an approximate value of the limiting current density is

$$i_l = 0.7\,nc \qquad (6.33)$$

Deposition of copper from a well-supported solution of 0.1 M $CuSO_4$ proceeds at a limiting current density on the order of 1 mA/cm² in an unstirred cell and 100 mA/cm² in a stirred cell. Limiting current density varies with electrode size, electrode orientation, electrolyte composition, temperature, hydrodynamics, and a variety of other factors. These two expressions were developed simply to illustrate the order of magnitude expected in a typical cell, but they should not be used for calculating limiting current densities; correlations for making accurate estimates of limiting current behavior are given below.

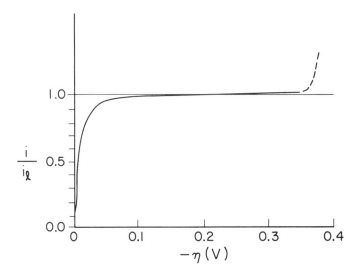

Figure 6.2: Schematic of overpotential vs. current density for the copper deposition reaction. The plateau represents the limiting current density. Hydrogen evolution (dashed line) commences at more negative potentials.

If current density is increased above the limiting current density, another reaction commences and the two reactions proceed simultaneously. For the case of deposition from acid-copper electrolyte, the reaction that proceeds above the limiting current density for copper deposition is the hydrogen evolution reaction. A plot of overpotential vs. current density appears in Fig. 6.2.

The use of supporting electrolyte can significantly reduce ohmic losses, but it generally reduces limiting current density as well. There are two factors that tend to reduce the rate of mass transport. Supporting electrolyte increases the viscosity of the solution and thereby decreases the maximum ion velocity. In addition, supporting electrolyte increases solution conductivity so that a smaller electric field is required to drive a specified current. The reduced magnitude of the electric field reduces the migration component of ion transport.

An important exception to this generalization occurs when the effects of migration tend to reduce ion flux. When a negatively charged species is reduced at a cathode, the electric field tends to retard the flux of that species. For example, ferricyanide can be reduced to ferrocyanide ion at a cathode. Consequently, reduction of the field through the addition of supporting electrolyte increases the limiting current density.

The ratio of limiting current with no supporting electrolyte to that with a large excess of supporting electrolyte is about 2. This ratio varies to some extent with the nature of the components and the mass transport conditions. For a number of metal deposition reactions on a rotating disk electrode, Newman [1] has shown that the results are similar. We define r as the ratio of supporting electrolyte to total electrolyte. A plot of the limiting current at a specified supporting electrolyte concentration to that where a large excess is present, $i_l(r)/i_l(r = 1)$ vs. $\sqrt{r}$ shows the decrease in limiting current density with the increase of supporting electrolyte concentration (Fig. 6.3). For this plot total dissociation of the ionic components was assumed.

In the case of a redox couple, the limiting current density is a much weaker function of r. In the metal deposition case, ions are only consumed at the cathode; by contrast, ions are both consumed and produced at the cathode for a redox couple. Consequently, the conductivity changes are less pronounced for the redox couple, and the addition of supporting electrolyte is less influential.

6.3 Concentration Overpotential

Overpotential associated with mass transport limitations is called concentration overpotential. It results primarily from the concentration cell that is established between the bulk electrolyte and the electrode surface. A second contribution results from the diffusion potential, which is calculated by solving Eq. 6.7 for the following potential gradient.

$$\nabla \phi = -\frac{i}{\kappa} - \frac{F}{\kappa} \sum_i z_i D_i \nabla c_i \qquad (6.34)$$

The first term on the right-hand side represents the potential gradient resulting from the ohmic drop. The second represents the contribu-

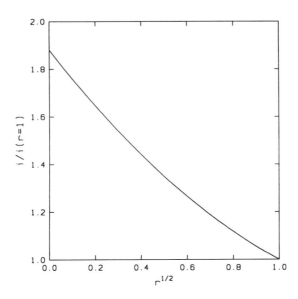

Figure 6.3: Effect of supporting electrolyte on limiting current density for $CuSO_4$-H_2SO_4 electrolyte on a rotating disk electrode.

tion from concentration gradients, and the resulting potential drop is known as the diffusion potential. Concentration overpotential results from both the concentration cell term and diffusion potential. The latter term is evaluated by integrating the concentration gradient over the diffusion layer. If we assume that conductivity variations in the diffusion layer are small, then the concentration overpotential is

$$\eta_c \;=\; \frac{RT}{nF} \ln \frac{c_{i0}}{c_{i\infty}} \;-\; \frac{F}{\kappa} \int_0^\delta \sum_i z_i D_i \frac{\left(c_{i\infty} - c_{i0}\right)}{\delta}\, dx \qquad (6.35)$$

In this equation we have used the Nernst diffusion layer model.

The assumption of small conductivity variations is most appropriate when current density is small or when there is an excess of supporting electrolyte. In the latter case the value of the conductivity in the denominator of the second term on the right-hand side is large, the diffusion potential is small, and Eq. 6.35 reduces to

$$\eta_c = \frac{RT}{nF} \ln \frac{c_{i0}}{c_{i\infty}} \tag{6.36}$$

where the subscript i refers to copper ion concentration. The difficulty in using any of the above equations for concentration overpotential is that the surface concentration c_0 is not readily determined. By dividing Eq. 6.30 by Eq. 6.31, we obtain a relation between surface concentration and local current density.

$$\frac{i}{i_l} = \frac{(c_\infty - c_0)}{c_\infty} \tag{6.37}$$

Substitution of this expression into Eq. 6.36 yields

$$\eta_c = \frac{RT}{nF} \ln \left(1 - \frac{i}{i_l} \right) \tag{6.38}$$

Concentration overpotential at the cathode is always a negative quantity. The conventions for surface overpotential are also used here. The total overpotential is the sum of the surface and concentration terms. If we consider concentration overpotential at the anode, we cannot apply Eq. 6.38 directly because there is no analog of a limiting current density at that electrode. For a copper anode the current density increases until a film forms on the electrode and the current is sharply reduced. The concept of a Nernst diffusion layer can still be applied. Division of Eq. 6.30 by Eq. 6.31 and substitution into Eq. 6.36 yields

$$\eta_c = \frac{RT}{nF} \ln \left(1 + \frac{i\delta}{nFDc_\infty} \right) \tag{6.39}$$

Eqs. 6.38 and 6.39 require a knowledge of the limiting current density or Nernst diffusion layer thickness. Techniques for estimating these quantities are discussed below.

6.4 Calculating Limiting Current Density

In ordinary chemical systems the rate of mass transport is characterized by a mass transfer coefficient k_m, which is the proportionality constant relating the flux to the concentration driving force.

$$N_i = k_m(c_\infty - c_o) \tag{6.40}$$

In most systems the mass transfer coefficient characterizes the rate of transport due to convection and diffusion. Rigorous solutions to the pertinent transport equations are rarely available except in special cases. In highly symmetric systems where the fluid flow is laminar, satisfactory analytical solutions yield expressions for the mass transfer coefficient. For example, the rate of mass transfer to a flat plate or rotating disk electrode can be calculated from fundamental conservation equations. When the system does not display a high degree of symmetry and when the flow is not laminar, analytical expressions for the mass transfer coefficients cannot be obtained. We must then resort to a correlation.

Because mass transport in chemical systems has been studied extensively, hundreds of correlations are available in the literature. They are often cast in terms of dimensionless numbers, where the dimensionless flux is a function of the dimensionless fluid velocity multiplied by a dimensionless number characterizing the fluid properties. For example, in forced convection the correlation is often of the form

$$Sh = f(Re, Sc) \tag{6.41}$$

where Sh is the Sherwood number, Re is the Reynolds number, and Sc is the Schmidt number defined by

$$Sh = \frac{k_m L}{D} \tag{6.42}$$

$$Re = \frac{Lv}{\nu} \tag{6.43}$$

$$Sc = \frac{\nu}{D} \tag{6.44}$$

In these expressions ν is the kinematic viscosity and L is a characteristic length. Some authors use the Nusselt number in place of the Sherwood number, but the former is more commonly applied to analogous heat transfer correlations.

At the limiting current density an electrochemical system is under mass transport control. Because the limiting current density is

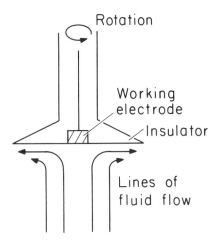

Figure 6.4: Rotating disk electrode schematic. A cylindrical electrode is embedded in an insulating disk. Fluid is pumped toward the electrode by rotation.

directly proportional to the flux, we can relate i_l to the Sherwood number. In an ordinary chemical system we can relate the flux to the mass transfer coefficient through Eq. 6.40. A surface concentration of zero corresponds to the limiting current condition, and the mass transfer coefficient can be expressed as

$$k_m = \frac{N_i}{c_\infty} \qquad (6.45)$$

The limiting current density in terms of the flux is

$$i_l = nFN_i \qquad (6.46)$$

Substituting Eqs. 6.45 and 6.46 into Eq. 6.42 yields

$$Sh = \frac{i_l L}{nFDc_\infty} \qquad (6.47)$$

Many mass transfer correlations have been developed for electrochemical cells over a range of flow conditions. A few have been derived from fundamental transport and conservation equations; mass transport to a rotating disk electrode (Fig. 6.4) is an example of this type of correlation.

The rotating disk electrode is commonly used in electrode kinetics and mass transport studies because its characteristics, based on known hydrodynamics, are well understood (see Ch. 8). As the disk rotates, fluid containing reactants is uniformly drawn toward its surface so that the reaction rate (current distribution) is uniform at the limiting current on the electrode. Flow to the disk is stable and can be maintained in the laminar regime over a wide range of rotation rates when the Reynolds number is less than 2×10^5. For typical experiments this criterion allows laminar operation up to several thousand rpm. The expression for mass-transport-limited behavior is

$$Sh = 0.62\, Re^{1/2}\, Sc^{1/3} \qquad (6.48)$$

This relation is known as the Levich equation. The characteristic dimension is the disk radius r. When we substitute the definitions for the dimensionless numbers into this equation, we obtain

$$i_l = 0.62\, \frac{nFDc}{r}\, \left(\frac{r^2 \omega}{\nu}\right)^{1/2} \left(\frac{\nu}{D}\right)^{1/3} \qquad (6.49)$$

$$= 0.62\, nFD^{2/3}\, \nu^{-1/6}\, c\, \omega^{1/2} \qquad (6.50)$$

where ω is the rotation rate (rad/s) and c refers to the reactant concentration in the bulk electrolyte. The dependencies are most apparent in Eq. 6.50. When the conditions of the Levich equation are fulfilled, a plot of the square root of the rotation rate vs. the limiting current density should yield a straight line. This result has been verified in numerous experiments.

For forced convection, exponents on the Reynolds number typically range from 0.5 to 0.9, depending on the specific geometry and flow regime. A direct proportionality to bulk concentration and relatively weak dependencies on physical properties of the fluid are also characteristic of mass transport correlations.

For turbulent flow an analytical solution of the hydrodynamic equations is not possible. Instead, mass transfer data are correlated with the dimensionless variables; coefficients and exponents reflect a

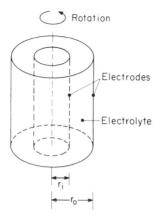

Figure 6.5: Rotating cylinder electrode schematic. Inner electrode rotates and creates a turbulent flow at moderate rotation rates.

fit of the data. A system invariably operated in the turbulent regime is the rotating cylinder electrode, shown in Fig. 6.5. Rotation of the inner cylinder at relatively low speeds (several hundred rpm) results in turbulent flow, and with proper design current distribution is uniform. The limiting current correlation developed by Eisenberg, Tobias, and Wilke [2] is

$$Sh = 0.791 \, Re^{0.7} \, Sc^{0.356} \qquad (6.51)$$

$$i_l = 0.791 \, \frac{nFDc}{d} \left(\frac{dv}{\nu}\right)^{0.7} \left(\frac{\nu}{D}\right)^{0.356} \qquad (6.52)$$

where d is the inner cylinder diameter and v is the surface velocity of the cylinder (cm/s). This expression is valid for $1000 < Re < 100{,}000$ and for $840 < Sc < 11{,}500$.

Both of the above correlations were developed for forced convection. Correlations are also available for free convection conditions.

In these systems fluid flow is driven by density differences created by reactions at the electrode surfaces. Consider a pair of vertical electrodes in initially quiescent electrolyte. If the processes occurring are metal deposition at the cathode and metal dissolution at the anode, then density gradients will be established as the electrode processes commence. At the cathode, solution immediately adjacent to the electrode will be depleted in metal ions and that region of the solution will become less dense. Metal dissolution will cause an increase in density of the solution adjacent to the anode. These gradients will drive the flow of solution.

The resultant flow will cause the limiting current density to increase substantially compared with a purely diffusive current. Flow due to density differences is usually characterized by the dimensionless Grashof number,

$$Gr = \frac{g(\rho_\infty - \rho_0)L^3}{\rho_\infty \nu^2} \tag{6.53}$$

where g is the acceleration due to gravity (980 cm/s^2), ρ_∞ is the bulk solution density, and ρ_0 is the density of solution immediately adjacent to the electrode. When operating at the limiting current density, the fluid density at the electrode surface can be taken as that of pure solvent.

Operation at the limiting current density may result in laminar or turbulent flow. When the product of the Grashof and Schmidt numbers is between 10^4 and 10^{12}, the flow is laminar and the following correlation is applicable [3]:

$$Sh_{avg} = 0.66 \, (ScGr)^{0.25} \tag{6.54}$$

When the condition $4 \times 10^{13} < ScGr < 10^{15}$ is met, the flow is turbulent and the following correlation is appropriate for vertical electrodes:

$$Sh_{avg} = 0.31 \, (ScGr)^{0.28} \tag{6.55}$$

The characteristic length is the electrode length. These correlations are valid for Schmidt numbers near 1000, which is the case for many common liquid electrolytes.

Correlations are available for standard, symmetric cell geometries (plates, cylinders, spheres, cones, etc.) under a wide variety of hydrodynamic conditions. A comprehensive list of correlations is presented by Selman and Tobias [4], a number of which are tabulated in Appendix E.

6.5 Unsteady-state Behavior

The treatment so far is relevant to the steady-state operation of electrochemical systems. Because most industrial systems operate at steady state, those relations are most useful. There are certain types of operations that are inherently unsteady state. For example, in a pulse-plating operation, the current is turned off periodically for short intervals to allow the concentration of metal ions to increase at the electrode surface. Just after the current is turned on again, the limiting current density is larger than in the steady-state case. After a certain time the boundary layer grows until it approaches the steady-state value, when the current is again turned off. Proper regulation of the on-off periods permits operation at relatively low fractions of the limiting current density, which is often useful in producing an improved deposit morphology. Several analytical techniques also rely on unsteady-state behavior. Both kinetic and mass transfer parameters have been calculated from experiments based on unsteady-state measurements.

Several cases of unsteady-state behavior have been studied, and analytical solutions have been derived for several of these cases. Just after applying a specified current or potential to an electrode but before significant density differences have been established, convective motion can be ignored. If this approximation is taken to be valid over a time interval of interest, then the governing differential equation is Fick's second law, or

$$\frac{\partial c}{\partial t} = D \frac{\partial^2 c}{\partial x^2} \tag{6.56}$$

Use of this equation also implies that the diffusivity is assumed to be constant. Constant current or potential operation can be reflected in the boundary conditions.

We first consider the constant current (galvanostatic) case. Before the current is switched on, the current, which is proportional to the potential gradient, is zero for all x. After the current is switched on, the concentration gradient at the surface becomes a constant value, depending on the current density; in well-supported electrolyte the proportionality between the current density and concentration gradient is expressed in Eq. 6.29. In this discussion we consider only cathodic processes and treat the current density as a positive quantity. The corresponding boundary conditions are

$$t = 0 \qquad \frac{\partial c}{\partial x} = 0 \qquad (6.57)$$

$$t > 0 \qquad \frac{\partial c}{\partial x} = \frac{i}{nFD} \qquad (6.58)$$

$$c(\infty, t) \qquad c_\infty \qquad\qquad (6.59)$$

Eq. 6.56 is a linear partial differential equation, which can be solved by several techniques, including separation of variables and the Laplace transform method. The solution to this equation was first presented by Sand [5].

$$c_\infty - c_0 = \frac{2i}{nF} \left(\frac{t}{\pi D} \right)^{1/2} \qquad (6.60)$$

$$= \frac{1.13\,i}{nF} \left(\frac{t}{D} \right)^{1/2} \qquad (6.61)$$

The concentration at the electrode interface drops in proportion to the square root of time. This equation indicates that the interfacial concentration will reach zero for any specified current. In reality, if the current is sufficiently small, convective forces will play a larger role at longer times, and the original assumptions will not be valid. Under these conditions steady state may be achieved before the surface concentration falls to zero. If the current is sufficiently high, the concentration at the interface will reach zero at a time referred to as the transition time (τ). With c_0 equal to zero, Eq. 6.61 becomes

$$\tau = \frac{\pi D}{4} \left(\frac{nFc_\infty}{i} \right)^2 \qquad (6.62)$$

In terms of current density, this equation becomes

$$i = \frac{nFc_\infty}{1.13} \left(\frac{D}{\tau} \right)^{1/2} \qquad (6.63)$$

In different experiments in the same system, the product of the current density and the square root of the transition time are constant

$$i\sqrt{\tau} = C_1 \qquad (6.64)$$

Eq. 6.60 is a special case of the more general solution, which gives the entire concentration profile. It can be obtained by using the Laplace transform method.

$$
\begin{aligned}
c_\infty - c_0 = \frac{i}{nFD^{1/2}} \Bigg[\left(\frac{4t}{\pi}\right)^{1/2} \exp\left(-\frac{x^2}{4Dt}\right) \\
- \frac{x}{D^{1/2}} \mathrm{erfc}\left(\frac{x^2}{4Dt}\right)^{1/2} \Bigg]
\end{aligned}
\qquad (6.65)
$$

where erfc (y) is the error function complement,

$$\mathrm{erfc}\,(y) = 1 - \mathrm{erf}\,(y) \qquad (6.66)$$

and erf(y), the error function, is the commonly tabulated function

$$\mathrm{erf}\,(y) = \frac{2}{\sqrt{\pi}} \int_0^y e^{-v^2}\, dv \qquad (6.67)$$

A plot of the concentration distributions for several values of dimensionless time (t/τ) appears in Fig. 6.6. Values chosen for the parameters were $n = 1$, $D = 10^{-5}$, and $c_0 = 0.1$ M. Because the current density was constant, the concentration gradients at the electrode surface were all equal. Combining Eqs. 6.37, 6.61, and 6.63 gives an expression for the concentration overpotential as a function of time.

$$\eta_c = \frac{RT}{nF} \ln\left(1 - \sqrt{\frac{t}{\tau}}\right) \qquad (6.68)$$

The Nernst diffusion layer thickness at the transition time can be calculated from Eq. 6.30.

$$\delta_\tau = \frac{nFDc_\infty}{i} \qquad (6.69)$$

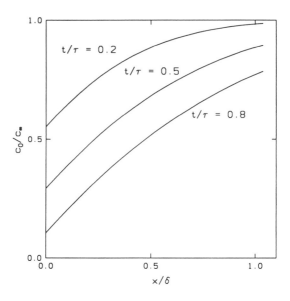

Figure 6.6: Concentration vs. distance for several values of dimensionless time under galvanostatic conditions.

At the transition time the surface concentration is zero, the current density is given by Eq. 6.63, and the equivalent film thickness in terms of the transition time is

$$\delta_\tau \; = \; 1.13 \, \sqrt{D\tau} \tag{6.70}$$

Under the conditions specified for Fig. 6.6, τ is about 7 s and δ_τ is about 10^{-2} cm.

The potentiostatic (constant potential) case can also be treated with analytical methods. The governing equation is Eq. 6.56, but the boundary conditions need to be recast. Because we are neglecting kinetic effects, all of the overpotential is related to mass transport limitations. From Eq. 6.36 we see that a fixed value of concentration overpotential implies that the surface concentration is also constant. Before the cell is stepped to a fixed potential, the concentration gradients are zero, and the boundary conditions become

$$t \; = \; 0 \qquad \frac{\partial c}{\partial x} \; = \; 0 \tag{6.71}$$

$$t > 0 \qquad c_0 = C_1 \qquad (6.72)$$

$$c(\infty, t) \qquad c_\infty \qquad (6.73)$$

The solution to the unsteady-state equation with these boundary conditions is

$$i = nF(c_\infty - c_0) \left(\frac{D}{\pi t}\right)^{1/2} \qquad (6.74)$$

This equation is referred to as the Cottrell or Cottrell-Stephan equation. If the potential is stepped to a value where the surface concentration is zero, then all of the variables except the diffusivity in the equation can be evaluated. A plot of current density vs. the inverse of the square root of time should yield a straight line from which the diffusivity can be calculated. Diffusion control can usually be maintained on a shielded, planar electrode for periods on the order of 10 s.

The above examples illustrate just a few of the many solutions available to the unsteady-state equation. Additional solutions for other boundary conditions in planar geometries and for other coordinate systems are also available [6].

6.6 Problems

1. Copper is being deposited from a solution that is 1 M $CuSO_4$ and 2 M H_2SO_4 onto a planar electrode. Sketch the profile of each ionic component near the electrode. Determine the direction of the flux due to migration and diffusion for each species. Note that the net flux at the electrode must be zero for non-reacting species. Also, electroneutrality must be maintained in the diffusion layer.

2. For the cathodic deposition of copper from 0.5 M $CuSO_4$ and 0.5 M H_2SO_4 electrolyte, the kinetic parameters are $\alpha_c = 0.5$ and $i_0 = 1$ mA/cm^2.
a) Calculate the current density at which hydrogen evolution may be expected if only kinetic limitations are taken into account.
b) Make the best estimate of the limiting current density if two plane parallel copper electrodes, 2-cm long are immersed in a beaker of unstirred electrolyte.

c) Estimate the thickness of the Nernst diffusion layer. Compare this value with the value incorporated into Eq. 6.32.

d) Estimate the current density at which hydrogen will evolve if both kinetic and mass transport effects are taken into account.

3. The rate of corrosion is sometimes limited by the mass transport rate of the oxidant, frequently hydrogen ions or molecular oxygen. For a steel pipe with seawater flowing at 2 m/s, the corrosion rate is limited by the rate of oxygen mass transfer. The oxygen content at 25°C is 8 ppm.

a) Calculate the maximum flux of oxygen to the surface in a 2.5-cm diameter pipe at a point 20 cm from the entrance.

b) Estimate the limiting current density at the same point.

c) Estimate the corrosion rate 20 cm down the pipe in mm/y.

d) Calculate the limiting current density on a 2.5-cm cylindrical electrode rotating in the same electrolyte with a surface velocity of 2 m/s.

e) Calculate the rotation rate on a cylinder corresponding to the limiting current density at a point 20 cm down the pipe.

4. An electrochemical cell consists of two concentric cylindrical electrodes. Both are copper and the inner cylinder (cathode) rotates at 500 rpm. The inner cylinder radius is 1.27 cm, the outer is 6.8 cm, and the height is 14.8 cm. The electrolyte is 0.6 M $CuSO_4$ and 1 M H_2SO_4 at 25°C. The applied potential is 1.4V.

a) Calculate the current density if only ohmic drop is considered.

b) Estimate the current density if kinetic, ohmic, and mass transport effects are taken into account. Because several terms in the voltage balance are nonlinear, a simple computer program might be appropriate for the problem solution.

c) Determine the concentration overpotential at the cathode.

d) Calculate the thickness of the deposit after two hours.

5. Metal is being removed from acid electrolyte in a flow channel at the limiting current density. The cathode is 10-cm long and the electrode separation is 5 cm. Under laminar flow conditions at

a Reynolds number of 800, the average limiting current density is 10 mA/cm².

a) Calculate the average limiting current density when the flow rate is doubled.

b) Would the total metal deposited be greater or less if the 10-cm electrodes were cut into five 2-cm sections and distributed down the channel as a series of segmented electrodes operated at the limiting current density? Explain your answer in qualitative terms.

6. McDonald et al. [*J. Phys. Chem.*, **90**, 196 (1986)] measured the diffusion coefficient of cupric ions in near-critical aqueous solutions using an unsteady-state method. They stepped the potential to a sufficiently negative value, where the concentration of ions at the electrode surface was assumed to be zero. At 245°C in 8.0 mM $CuSO_4$ they obtained the following data on a planar electrode 2 mm in diameter:

t (ms)	2	3.3	6.3	8.2
I (mA)	10	9	6	5

Estimate the diffusion coefficient for the cupric ion under these conditions.

Bibliography

[1] J. S. Newman, *Int. J. of Heat and Mass Transfer*, **10**, 983 (1967).

[2] M. Eisenberg, C. W. Tobias, and C. R. Wilke, *J. Electrochem. Soc.*, **101**, 306 (1954).

[3] N. Ibl, *Electrochim. Acta*, **1**, 117 (1959).

[4] J. R. Selman and C. W. Tobias, *Advances in Chemical Engineering*, **10**, 212 (1978).

[5] H. J. S. Sand, *Phil. Mag.*, **1**, 45 (1900).

[6] J. Crank, *The Mathematics of Diffusion* (Oxford: Clarendon Press, 1975).

Chapter 7

Modeling and Simulation

The motivation for developing a mathematical model of an electro-chemical process is to reduce the experimental effort required to design or optimize the process. For a system of any complexity, it is invari-ably more cost-effective to perform a series of computer experiments than to perform the same experiments in the field. Once developed, a model can be refined through comparison with laboratory and field data and become more useful as a design tool.

7.1 General Modeling Considerations

When modeling an electrochemical process, we are interested in re-taining the essential features of the system while avoiding unnecessary mathematical complexity. Simplification of a model is accomplished by making assumptions that are appropriate for the degree of accuracy that is required. In some cases the validity of assumptions made can be checked through an analysis of results; in other cases, the validity of assumptions made relies more on the judgment of the engineer.

The diversity of electrochemical processes is such that no general model can cover most systems of interest. Instead, a classification system has arisen so that many electrochemical processes have been grouped, and generalizations regarding a particular classification have been developed. The simpler classes of models are similar to those from mathematical physics; consequently, several of these classes have been studied extensively over the last hundred years. In recent years there has been an increasing tendency to develop more accurate models. The impetus for this increasing sophistication has been the general access to computer facilities that allow the implementation of these models. Present simulations of industrial processes rely almost exclusively on computer calculations.

The essential features of most models rely on fundamental laws such as material and energy conservation as well as empirical laws such as the flux and rate expressions. A flux expression developed for dilute solutions and the fairly general Butler-Volmer rate expression (or simplified forms) are commonly used in general models.

In addition to a physicochemical description of a system, a model of the system's geometry must be developed. The complexity of a system can be reduced considerably if it can be adequately modeled in one or two dimensions. This modeling is most readily accomplished in systems displaying a high degree of symmetry. When concentration gradients must be taken into account, a model of fluid dynamics is also required.

Once a mathematical model of a system has been formulated, a method for performing a simulation must be chosen. Many models can be cast in forms amenable to analytical techniques have already been studied, and expressions or graphical descriptions of current and potential distributions are available in the literature. For more complex systems a numerical method is required to solve the pertinent equations. These techniques involve a division of the domain into subunits. Relationships between the subunits must be satisfied so that global conditions are simultaneously satisfied. This process is accomplished by postulating trial functions that are corrected at each iteration. If the process is stable, then a larger number of iterations leads to a more accurate solution.

The final step in the simulation process is to determine whether the results appear to be reasonable. Often the computer algorithms required for a solution are so complex that small errors in the computer programs are difficult to detect. Direct comparison of simulated results with experimental values provide the best check of a model. If serious discrepancies arise, the investigator must then decide whether there are bugs in the program or whether the model itself is inadequate. The former problem can often be resolved by using two different solution techniques. The latter problem is more difficult to resolve because it may result from neglecting important effects or from poor parameter estimation.

7.2 Model Classification

For the purposes of discussing the difficulty of performing a simulation from a model, we often classify a model with regard to its geometry and physicochemical effects. In the present discussion we will only consider electrodes where the reaction occurs along a non-porous surface at steady state. Models where concentration gradients are ignored are referred to as potential theory models. When concentration gradients are significant, their quantitative characterization relies on a description of the fluid motion; the fluid flow can be more difficult to simulate than the electrochemical system. These models are called convective-transport models.

Simulations are often classed in terms of the resulting current distribution:

Primary Current Distribution: Only electric field effects are considered.

Secondary Current Distribution: Kinetic and electric field effects are considered.

Tertiary Current Distribution: Concentration gradients, kinetic limitation, and electric field effects are considered.

Mass Transport-limited Current Distribution: Concentration gradients alone govern the distribution.

Actual systems contain all of the elements listed in the most general tertiary current distribution model. Several quantitative guides, as well as engineering judgment, are required to determine which factors can be approximated or neglected.

Primary and secondary current distributions constitute a class of potential theory models. The simplest model of an electrochemical system is the primary current distribution. The partial differential equation governing the potential distribution in solution was derived

in the previous chapter. In the absence of concentration gradients, we
have Laplace's equation (Eq. 6.10):

$$\nabla^2 \phi = 0 \qquad (7.1)$$

Because this equation also governs the conduction of heat in solids,
steady-state diffusion, and electrostatic fields, it has been studied in
great detail. For a primary current distribution we neglect overpotential at the electrode/electrolyte interface; therefore, the potential
immediately adjacent to the electrode surface is modeled as a constant
potential surface. The current density is proportional to the potential
gradient, as indicated by

$$\mathbf{i} = -\kappa \nabla \phi \qquad (7.2)$$

At an insulated surface, the current density and normal potential gradient are zero. If the assumptions of a primary current distribution
can be justified, it is usually a fairly straightforward matter to obtain
a solution to the governing equations. Mathematically, the problem
with only constant potential boundary conditions is called a Dirichlet
problem. Equation 7.1 has analogs in which temperature or concentration variables replace the potential variable. A large number of
solutions to the temperature analog are presented by Carslaw and
Jaeger [1].

In a secondary current distribution, we need to account for electrode kinetics. Because kinetic phenomena are occurring within a
few tens of ångströms of the electrode surface, it is legitimate to incorporate kinetic effects in the boundary conditions. For a secondary
current distribution model, Laplace's equation still governs the potential in solution, but the boundary conditions are no longer considered
to be constant value potentials. Instead, the potential near the surface is governed by the following current-overpotential (polarization)
expression:

$$\eta_s = f(i) \qquad (7.3)$$

Common models for the polarization expression include the Tafel and
linear equations.

The tertiary current distribution model is more general and more
difficult to deal with. Because we are concerned with concentration

variations, we need to understand system hydrodynamics as well. In only a few systems are the hydrodynamics sufficiently well described to allow a rigorous solution of the governing equations. One technique for simplifying the mathematics is to divide the system into two regions: a bulk electrolyte region, where the solution is well-stirred, and a boundary layer region, where the concentration gradients are most significant. In the bulk region concentration gradients are negligible, and Laplace's equation is valid. In the mass transport boundary layer region the convective-diffusion equation can be used to model ion transport. A matching condition at the bulk/boundary layer region is provided by requiring that the current density be equal at the interface.

For the mass transport-limited case, convection and diffusion limit the overall rate of the system. In aqueous systems the convective-diffusion equation governs, and the hydrodynamic conditions strongly influence the current distribution on the electrode surfaces. The difficulty of solving the equations is strongly related to the difficulty of the hydrodynamic problem.

Systems that we model do not always fall neatly into one of these categories. Rather, they fall on a spectrum, and some judgment must be exercised in deciding how much error can be tolerated when an assumption is made to simplify a problem. In determining whether a primary current distribution adequately describes a system, we need to know the relative importance of the system kinetics with respect to the ohmic component. The dimensionless number that characterizes this ratio is the Wagner number,

$$Wa = \frac{\kappa \left(\frac{\partial \eta_s}{\partial i} \right)}{L} \tag{7.4}$$

where the partial derivative is evaluated at the average current density and L is a characteristic length. The partial derivative characterizes charge-transfer resistance and the inverse of the conductivity characterizes an ohmic resistance in solution. If the Wagner number is on the order of one, then the kinetic component is substantial, and a primary current distribution cannot accurately describe the system. If the Wagner number is small, then ohmic resistance dominates, and the use of the primary current distribution model is justified.

The characteristic dimension L is used to quantify the scale of a system. It should be chosen so that the current distribution is sensitive to changes in L. For some systems the proper choice of the characteristic length is an obvious one. In the case of a disk or cylinder electrode, the radius (or diameter) of the electrode is invariably used. In other systems the choice is not immediately apparent. For example, if an electrode surface were modeled as a sinusoidal function, amplitude, wavelength, or electrode separation could serve as a characteristic dimension. In general, the amplitude would be the most appropriate choice. Because the Wagner number is inversely proportional to L, increasing the amplitude generally reduces the magnitude of the Wagner number. A reduced Wagner number implies that the current distribution is more primary in character, and that the current distribution is less uniform, as discussed in the following section.

7.3 Current Distribution Characteristics

Based on the previously described classification system, we can make a number of generalizations regarding the characteristics of current and potential distributions. This information can be used to give an intuitive check of a simulation. In addition, we can determine which regions are subject to the largest potential gradients, where difficulties in carrying out a simulation may arise.

The primary current distribution model is an idealized one that provides mathematical simplicity. Its use is only justified in cases where the Wagner number is low. A characteristic of the primary current distribution is that it is always less uniform than a secondary current distribution in a specified system. The only exception is the case where the secondary current distribution is uniform for all values of the Wagner number, such as in a concentric cylinder system. Both the primary and secondary current distributions are uniform for the rotating cylinder electrode. The non-uniform nature of the primary current distribution is caused by the variation in electrolyte resistance for different current paths. Often, the minimum anode-cathode separation at a point on the working electrode gives a good indication of the ohmic resistance to current flow at that point; there are ex-

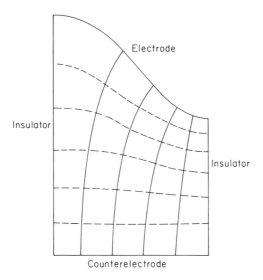

Figure 7.1: Equipotentials (dashed lines) with flux lines (solid lines) perpendicular to the equipotentials at all points.

ceptions to this generalization especially where electrode curvature or intersection with an insulator occurs.

A primary current distribution results from a solution of Laplace's equation. Because it is a linear equation with constant potential (or constant potential gradient) boundary conditions, the current density scales linearly. The relative current distribution is unaffected by changes when all system dimensions are changed by an arbitrary factor. This characteristic implies that once we have a current distribution for a given cell, we know the current distribution for a scaled-up version of the same cell.

Current flow is always perpendicular to equipotential surfaces in a primary current distribution model. Because there is no current flow to an insulated surface, equipotential surfaces—or lines, in the two-dimensional case—intersect insulated surfaces at right angles. Current flow is always normal to electrodes, which are equipotential surfaces. A two-dimensional plot of the equipotential and flux (current density) lines appears in Fig. 7.1. Equipotential lines tend to follow electrode surfaces near the electrodes. Because the flux lines must be perpendicular everywhere, current tends to concentrate near convex surfaces. In the extreme convex case, the electrode becomes a sharp peak, where the derivative is discontinuous, and the local current density becomes infinite. Although the current density becomes infinite, it does so along an infinitesimal line so that, in theory, the total current on the electrode remains finite. Near concave surfaces, local current density is reduced by the curvature.

Generalizations regarding current distribution at the intersection of an insulator and an electrode are possible for a primary current distribution. If a planar electrode intersects an insulator at an obtuse angle, the current density along the line of intersection is infinite, as illustrated in Fig. 7.2a. This description applies to an electrode that is coplanar with an insulator. Qualitatively, the reason for this phenomenon is that an equipotential line near the electrode is parallel to it. At the point of intersection the equipotential line must become perpendicular to the insulator. Because the equipotential line must change direction rapidly in the intersection region, the equipotential attains a sharply convex shape. The convex curvature results in concentration of the current at the intersection.

If the intersection of the electrode and the insulator takes place at an acute angle (Fig. 7.2b), the current density is zero at that intersection. The qualitative reason is the same as in the obtuse case, but for the acute angle the equipotential line becomes concave; therefore, current is directed away from the line of intersection.

It is only when an insulator intersects an electrode at a right angle (Fig. 7.2c) that current density becomes finite. In this case an equipotential line near the electrode automatically satisfies the conditions of being parallel to the electrode and perpendicular to the insulator. No abrupt bending of the equipotential line is required, and a uniform primary current distribution is readily arranged in a properly designed cell. Because a uniform current distribution is most appropriate for conductivity measurements, conductivity cells are typically constructed from two, parallel, circular electrodes inside a glass cylinder. In this case the electrodes are perpendicular to the insulating surfaces, and the primary (and secondary) current distributions are uniform.

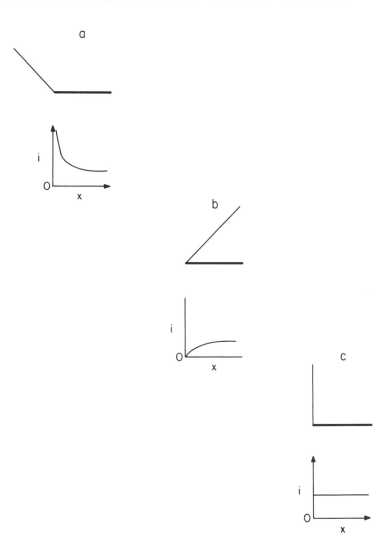

Figure 7.2: Three configurations of electrode-insulator intersections with corresponding primary current distributions. Horizontal lines represent electrodes and intersecting lines represent insulators. a) Obtuse angle; b) Acute angle; c) Right angle.

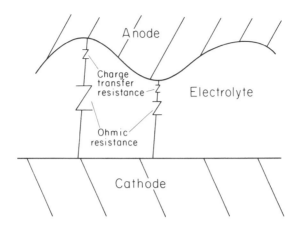

Figure 7.3: Electrical analogy of ohmic and charge-transfer resistance. As charge-transfer resistance increases, total resistance and current along each current path becomes more uniform.

Inclusion of electrode kinetics in a model tends to even out the current distribution compared to the primary case. This effect can be thought of in terms of the additional charge-transfer resistance, which acts in series with the ohmic resistance (Fig. 7.3). In this analogy as the resistor at the surface becomes larger, it tends to dominate the system. The total resistance in any current path tends to become constant. Because the applied potential between the electrodes is constant, the current density also tends toward a constant value. As the resistor representing ohmic losses becomes negligible compared to the resistor representing kinetic resistance, current distribution becomes uniform. This condition is characterized by a Wagner number approaching infinity. At small values of the Wagner number, a primary distribution model is appropriate, and at very high Wagner numbers, current distribution becomes uniform.

Although Laplace's equation still applies in the region representing bulk electrolyte, the boundary conditions no longer reflect constant potential values. Scaling a system must now be considered in terms of the Wagner number. We can make several observations regarding the current distribution based on the behavior of the Wagner number. There are three factors in the Wagner number. Increasing the characteristic dimension reduces the Wagner number and causes the current distribution to be less uniform. Increasing electrolyte conductivity (reducing the resistivity) tends to make a current distribution more uniform. The derivative in Eq. 7.4 represents the slope of the polarization curve evaluated at the average current density. A steeper slope is characteristic of a larger kinetic resistance and a more uniform current distribution. The derivative is a constant only for a linear overpotential model.

For a linear current density-overpotential equation of the form

$$\eta_s = \beta i \tag{7.5}$$

the Wagner number is

$$Wa = \frac{\kappa \beta}{L} \tag{7.6}$$

Because both the ohmic drop and overpotential vary linearly with current density, its effect cancels, and the Wagner number is independent of current density.

For the Tafel model:

$$\eta_s = B \log \frac{i}{i_0} \tag{7.7}$$

the Wagner number is

$$Wa = \frac{\kappa B}{Li} \tag{7.8}$$

where the current density is evaluated at the average current density. Because kinetic resistance varies logarithmically with current density—while ohmic resistance varies linearly—the Wagner number becomes a function of current density. When all other factors are held constant, a higher current density leads to a less uniform current distribution.

As discussed in Chapter 5, if the linear and Tafel models are based on an analysis of a single, elementary, charge transfer reaction, then the constants β and B have the following values:

$$B = \frac{2.303 \, RT}{\alpha F} \tag{7.9}$$

$$\beta = \frac{RT}{i_0(\alpha_a + \alpha_c)F} \tag{7.10}$$

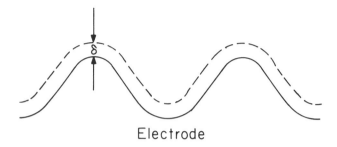

Figure 7.4: Example of a macroprofile, where the mass transfer boundary layer (dashed line) tends to follow the profile.

In other cases the same forms (Tafel or linear) can be used and the transfer coefficients treated as fitting parameters.

From these relations we can determine the effect of changes in temperature on the Wagner number. When operating in the Tafel region at constant current, both conductivity and B increase with temperature; therefore, the Wagner number also increases. In the linear region, exchange current density increases more rapidly than the linear function in the denominator so that β increases with increasing temperature. In this case the Wagner number increases with increasing temperature.

When using a tertiary current distribution model, concentration overpotential is an additional resistance (a mass transport resistance) acting in series at the electrode surface. Because the location of significant reactant depletion at the electrode surface does not necessarily correspond to the high current density region, surface overpotential and concentration overpotential do not necessarily reinforce each other. With few exceptions a secondary current distribution is more uniform than a primary current distribution, but a similar statement cannot be made when comparing the tertiary and primary cases. Because hydrodynamics and geometry play a large role in determining mass transport behavior, interaction among these factors needs to be considered. For example, the mass transport in a relatively open recess, as shown in Fig. 7.4, can be enhanced by increasing fluid agitation. We would expect the diffusion layer thickness to be

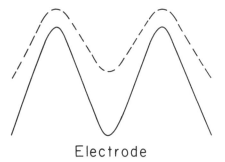

Electrode

Figure 7.5: Example of a microprofile, where the mass transfer boundary layer (dashed line) thickness is of the same order of magnitude as the dimension of the surface features. Boundary layer thickness varies.

fairly uniform over the entire profile. Concentration overpotential and surface overpotential would reinforce each other. In a more closed geometry as shown in Fig. 7.5, the diffusion layer would not tend to follow the electrode in the recess, and the region of most severe mass transport limitation would occur at the base of the recess, where the surface overpotential would be relatively low. In this type of geometry, sometimes called a microprofile, the concentration overpotential would tend to reduce the current density in the base of the recess, and cause the current distribution to be less uniform than the secondary distribution.

The dependence of concentration overpotential on current density is significantly different from that of surface overpotential. At low fractions of limiting current density, the slope of the concentration overpotential-current density curve is relatively small compared to the slope of the typical surface overpotential-current density curve. At high fractions of limiting current density, the opposite is true. For example, if the current density is 50 mA/cm² and the limiting current density is 100 mA/cm², then a typical value for the magnitude of the slope of the concentration overpotential-current density curve is

$$\frac{\partial \eta_c}{\partial i} = \frac{RT/nF}{i_l - i} \tag{7.11}$$

$$\frac{\partial \eta_c}{\partial i} = \frac{0.013}{0.050}$$
$$= 0.26 \text{ V/A}$$

For comparison we will consider a typical surface overpotential expression given by the Tafel equation.

$$\eta_s = B \log \frac{i}{i_0} \tag{7.12}$$

A typical value for B is 0.1 V, but the slope of the overpotential-current density curve is not a function of the exchange current density.

$$\frac{\partial \eta_s}{\partial i} = B/i \tag{7.13}$$
$$= 0.1/0.05 \tag{7.14}$$
$$= 2 \text{ V/A} \tag{7.15}$$

If we now make the same comparison at 95 mA/cm^2 (95% of the limiting current density) the slope of the concentration overpotential curve is 2.6, and the slope of the surface overpotential curve is 1.05. Because of the functionality of the kinetic expression, it is easier to characterize the current distribution with an average value of the current density. For the tertiary current distribution it is more appropriate to use the maximum current density when making an analogous Wagner number calculation.

7.4   Electrochemical Cell Simulation

Once we have chosen an appropriate model for an electrochemical system, we are interested in casting the pertinent variables in a mathematical form and finding a solution to the system of equations describing the system. Only the simplest cells can be described with a set of equations that can be solved by elementary means. In one-dimensional systems we may be able to reduce the problem to an ordinary differential equation, otherwise we must deal with partial differential equations. Those problems that we can reduce to one dimension include symmetric geometries such as parallel electrodes and concentric cylinders. With the simplest boundary conditions (constant

potential or constant potential gradient), many two-dimensional geometric arrangements have been treated [1] for analogous heat transfer
cases, i.e., those corresponding to the primary current distribution.
Extensions that include linear polarization were developed by Kasper
[2]. Analogs to the primary current distribution problem abound in
the literature. For symmetric cells it is likely that a solution to the
primary current distribution problem is available in the literature,
though not necessarily in electrochemical literature.

For systems not displaying a high degree of symmetry, a numerical
technique is usually required to solve the equations. Even in symmetric cells other complications, such as non-isothermal conditions or
complex kinetic expressions, can require the use of a numerical technique in obtaining a solution. Finite difference, finite element, and
boundary element techniques have been applied in the simulation of
electrochemical systems. Essentially any two- or three-dimensional
tertiary current distribution problem must be treated with a numerical technique; a survey of numerical solutions applied to electrochemical systems has been published [3].

In the above discussion we have limited ourselves to electrodes
where the charge transfer occurs at a smooth, planar or curved surface. Another important type of electrode is that where the charge
transfer occurs within the electrode. This type of porous electrode
system offers a large surface area per unit volume. In such systems a
significant portion of the charge transfer occurs within the electrode.
Because effective porous electrode systems require a network of small
diameter pores, diffusive processes become more important, and must
be treated differently. Several analytical solutions are available for
simplified systems, but again more realistic simulations require numerical techniques.

7.4.1 Analytical Techniques

The general mathematical techniques for solving resulting differential equations include direct integration, separation of variables, and
conformal mapping. A few, simple, one-dimensional problems can
be solved by direct integration. For example, the primary potential
distribution between two, infinite, parallel electrodes is simply ob-

tained by two integrations of Laplace's equation in one-dimensional, cartesian coordinates with constant value boundary conditions. The potential is a linear function of distance between the electrodes, and the current density is the same on both electrodes. Expressed mathematically,

$$\phi = V_c + \frac{(V_a - V_c)x}{L} \qquad (7.16)$$

where V_a and V_c are the anode and cathode potentials, and L is the electrode separation. The primary current density is

$$i = \frac{-\kappa (V_a - V_c)}{L} \qquad (7.17)$$

The primary potential distribution for concentric cylinder electrodes can also be treated by elementary analytical techniques. There are only potential variations in the radial direction; the current density is given by Ohm's law.

$$i = -\kappa \frac{d\phi}{dr} \qquad (7.18)$$

The total current in the cell must be constant; at any radial distance the current is the current density multiplied by the normal area, or

$$I = 2\pi r H i \qquad (7.19)$$

where H is the height of the cylinders. Substitution of Eq. 7.19 into Eq. 7.18 and integration from the inner to the outer cylinder yields

$$\int_{\phi_i}^{\phi_o} d\phi = -\frac{I}{2\pi H \kappa} \int_{r_i}^{r_o} \frac{dr}{r}$$

$$\phi_o - \phi_i = -\frac{I}{2\pi H \kappa} \ln \frac{r_o}{r_i} \qquad (7.20)$$

This expression demonstrates that the potential varies logarithmically with radial distance. If the current flows from the outer to the inner cylinder, then it is negative in this coordinate system, and the potential difference on the left-hand side is positive. Although the current density is uniform on each cylinder, it is lower on the outer cylinder because the surface area is larger.

The secondary potential distribution is readily calculated in this type of cell. If the total current, the electrolyte conductivity, kinetic expressions, and cell dimensions are specified, then we can calculate the potential distribution directly. For example, let $r_i = 1$ cm, $r_o = 5$ cm, $H = 10$ cm, $\kappa = 0.1$ ohm^{-1}cm^{-1}, $I = 0.5$ A and $\eta_s = 0.1$ log $(i/10^{-5})$ for both electrode reactions. Because we know the area of each electrode, we can calculate the current density and use those values in the kinetic expressions. If we consider the inner electrode as the cathode, we have

$$\eta_{sa} = 0.22 \text{ V}$$
$$\eta_{sc} = -0.29 \text{ V}$$

The magnitude of the ohmic drop in solution is given by Eq. 7.20.

$$\Delta\phi_{ohm} = \frac{I}{2\pi H \kappa} \ln \frac{r_o}{r_i}$$
$$= 0.13 \text{ V}$$

The magnitude of the voltage losses in the cell is equal to the sum of the three components or 0.64 V. The plot of the potential distribution in Fig. 7.6 illustrates the logarithmic dependence of the potential variation in solution.

If a rotating cylinder electrode system is used to measure overpotential, we can calculate the ohmic drop in solution between the working electrode and reference electrode. This ease of calculation is one of the advantages of working in a well-characterized system. The functionality of the potential distribution also influences the placement of a reference electrode. In this system, placement of the reference electrode near the inner (working) electrode is undesirable because potential gradients are highest in that region. This characteristic implies that small errors in reference electrode placement result in large errors in potential drop. With conductive electrolyte, placing the reference electrode near the outer electrode is advantageous because ohmic error and shielding of the working electrode are minimized. However, if electrolyte conductivity is low, the potential drop to a reference electrode is large. Calculation of the overpotential results from the subtraction of two large numbers to obtain the relatively smaller over-

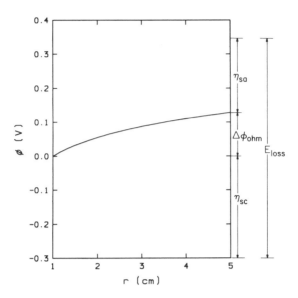

Figure 7.6: Potential distribution in a rotating cylinder system.

potential value. In this case placing the reference electrode closer to the inner electrode might be preferred.

Plane parallel electrodes and concentric cylinders are examples of systems that can be reduced to one dimension. In other systems the primary current distribution is not uniform, and a two- or three-dimensional simulation is required. An example of a two-dimensional simulation is the primary current distribution around parallel conducting cylinders (wires) as shown in Fig. 7.7. The potential distribution has been calculated by the method of images. This solution technique involves the conversion of the electric field into an equivalent field, where the potential is easier to calculate. The method is particularly useful for certain line, point, and plane electrode arrangements and is described in advanced texts on electricity and magnetism. The potential in solution for the parallel wire electrodes, shown in Fig. 7.8, is [4]

$$\phi = \frac{V}{2\ln\dfrac{d+p}{d-p}} \ln \left[\frac{2(d+p)(d+r\,cos\theta) + r^2 + p^2 - d^2}{2(d-p)(d+r\,cos\theta) + r^2 + p^2 - d^2} \right] \qquad (7.21)$$

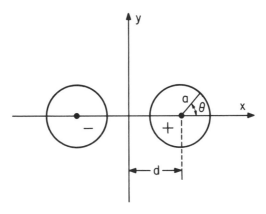

Figure 7.7: Parallel wire electrodes of radius a.

where d is half the distance between the centers of the wires, r the distance from the center of the right-hand wire to a point in solution, p is $\sqrt{d^2 - a^2}$, a is the radius of a wire, V is the applied potential (the potential of the right-hand wire is $V/2$ and the left-hand wire is $-V/2$), and θ is the angle between the x-axis (through the centers of wires) and the point of interest in solution. In this case equipotentials in solution are cylindrical, but are not coaxial with the conductors. The total current is

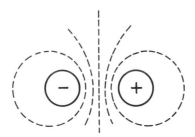

Figure 7.8: Equipotentials for parallel wire electrodes with a primary potential distribution.

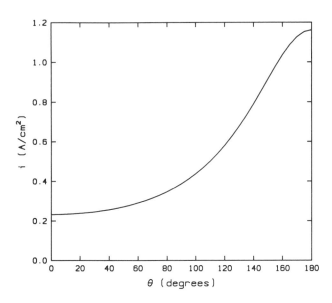

Figure 7.9: Primary current distribution on parallel wire electrodes: $V = 1$ volt, $a = 1$ cm, $d = 1.5$ cm, $\kappa = 1$ ohm^{-1}cm^{-1}.

$$I = \frac{2\pi l \kappa V}{\ln \dfrac{d + p}{d - p}} \qquad (7.22)$$

where l is the wire length. Alternatively, if current is specified, the above equation can be solved explicitly for the potential difference.

Local current density is proportional to potential gradient. Differentiating Eq. 7.21 with respect to r and evaluating the expression at $r = a$ gives the potential gradient. Multiplication by $-\kappa$ gives the current density as

$$i = \frac{a p \kappa V}{(d + p)(d - p)(d + a\ cos\theta)\ \ln \dfrac{d + p}{d - p}} \qquad (7.23)$$

A plot of the current distribution vs. θ appears in Fig. 7.9 for specified values of the geometric parameters. For comparison, note that infinite plane parallel electrodes placed at the distance of closest approach (1 cm) would have a current density of 1 A/cm^2 under the specified conditions. Because the wire electrodes are curved, current tends to be concentrated at the distance of closest approach, and the current density is about 16% higher than for the corresponding planar electrode case. For parallel wires the current distribution is the same provided that the ratio of a/d remains the same. As noted above, only changes in geometric parameters that are not proportional alter the primary current distribution.

For a primary current distribution we can identify the overall cell resistance from Ohm's law $V = IR$. For parallel wires, resistance is seen by an inspection of Eq. 7.22 to be

$$R = \frac{1}{2\pi l\kappa} \ln \frac{d+p}{d-p} \qquad (7.24)$$

Similarly, cell resistance for a concentric cylinder arrangement can be determined from Eq. 7.20.

$$R = \frac{1}{2\pi H\kappa} \ln \frac{r_o}{r_i} \qquad (7.25)$$

Potential distributions for many two-dimensional problems are expressed in terms of an infinite series. Often, obtaining a solution in terms of a series is most straightforward. Computer calculation is convenient even for slowly convergent series. Series involving special functions such as Bessel functions or Legendre polynomials are easily evaluated with standard software routines; however, some caution must be exercised when calculating the current distribution from a potential expressed in terms of a series. Because the current density is proportional to the normal derivative, the series must be both differentiable and convergent. A convergent series for the potential does not always lead to a convergent series for the potential gradient.

Consider the cell shown in Fig. 7.10. This illustration represents a cell with electrodes at right angles with small gaps at their junctions. The electrodes are infinite in the y-direction. By the principles of sectioning, we can place an insulator along the line at $x = H/2$. Using the method of separation of variables, we obtain the solution to Laplace's equation [5]:

$$\phi = \frac{4V_0}{\pi} \sum_{n=1,3,\ldots}^{\infty} \frac{1}{n} \exp\left(\frac{-n\pi}{H}y\right) \sin\left(\frac{n\pi}{H}x\right) \qquad (7.26)$$

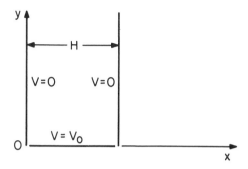

Figure 7.10: The schematic of a cell with perpendicular electrodes. Vertical electrodes are infinite and held at a potential of 0 volts. Horizontal electrode is held at V_0 volts.

To calculate a current distribution on a horizontal electrode, we need to differentiate the potential with respect to y and evaluate it at $y = 0$. When we perform this calculation term-by-term, we obtain

$$\frac{\partial \phi}{\partial y} = -\frac{4V_0}{H} \sum_{n=1,3,\dots}^{\infty} \sin \frac{n\pi}{H} x \qquad (7.27)$$

When we attempt to evaluate the derivative, we see that the series is simply the sum of sine terms, which never converges. Although the potential distribution is clearly convergent and yields valid results, current density cannot be calculated from the potential gradient evaluated from the series.

In this case an alternate solution is available. We can sometimes derive a solution to a primary current distribution problem based on conjugate functions. For the perpendicular electrode problem, the potential is [5]

$$\phi = \frac{2V_0}{\pi} \tan^{-1} \left[\frac{\sin(\pi x/H)}{\sinh(\pi y/H)} \right] \qquad (7.28)$$

If we differentiate the potential with respect to y and evaluate the function at $y = 0$, we obtain the potential gradient in the y-direction.

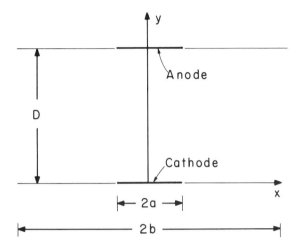

Figure 7.11: Schematic of plane electrodes embedded in insulators.

Multiplication by κ gives the magnitude of the current density as

$$i_y = \frac{2\kappa V_0}{H \, \sin(\pi x/H)} \tag{7.29}$$

Although there are a large number of analytical solutions corresponding to the primary current distribution, there are far fewer analytical solutions for secondary current distribution problems. Published solutions are often restricted to the simplest polarization equation (usually linear), and they are often restricted to a range of geometric parameters.

Some examples of analytical solutions to primary and secondary current distribution problems were presented by Wagner [6]. One of the geometric arrangements he studied was of plane parallel electrodes embedded in insulated walls as shown in Fig. 7.11. For the case where the counterelectrode is far removed from the working electrode, the primary current distribution, determined from conformal mapping, is

$$\frac{i}{i_{avg}} = \frac{2}{\pi} \frac{1}{[1 - (x/a)^2]^{1/2}} \tag{7.30}$$

Using a different technique (superposition of particular solutions), Wagner also solved the secondary current distribution problem for linear polarization as

$$\frac{i}{i_{avg}} = \frac{g(z)}{\int_0^1 g(z)dz} \tag{7.31}$$

where

$$g(z) = \{1 + \frac{a}{\pi\kappa\beta}[2 - (1+z)\ln(1+z)$$
$$-(1-z)\ln(1-z)]\}^{-1} \tag{7.32}$$

and

$$z = x/a \tag{7.33}$$

In these expressions β is the slope of the linear polarization curve. Because of approximations introduced, Eq. 7.31 is only valid for a certain range of Wagner numbers ($\kappa\beta/a > 0.6$). All of the plane parallel electrode solutions are restricted to $D >> 2a$ and $b >> a$. Plots of the current distribution for several values of the Wagner number appear in Fig. 7.12. The contrasting characteristics of primary and secondary current distributions are apparent in this figure. Where the electrode and insulating walls are coplanar, the current density becomes infinite for the primary current distribution. At higher values of the Wagner number, the current distribution becomes more uniform.

The examples cited above are just a few representative cases of analytical solutions. There are 11 natural coordinate systems in which Laplace's equation can be separated [7]. Our examples illustrate only solutions in cartesian and polar coordinates. The primary current distribution around an insulating sphere (bubble) adjacent to a planar electrode was solved in tangent-sphere coordinates by Sides and Tobias [8]. Newman [9] determined the primary current distribution on a disk electrode from a solution to Laplace's equation in rotational-elliptic coordinates.

7.4.2   Sectioning

In the numerical simulation of an electrochemical system, it is advantageous to perform the simulation on the smallest domain that

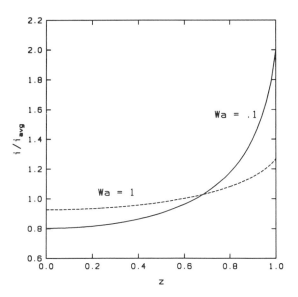

Figure 7.12: Secondary current distribution on a planar electrode embedded in an insulator. Current distribution is more uniform at higher Wagner numbers.

accurately describes the behavior of the cell. Often the domain can be reduced by taking system symmetry into consideration. Because both geometric and field symmetry must be taken into account, only potential theory problems are generally amenable to a reduction in domain size.

The principles governing reduction of the domain from symmetry considerations (sectioning) can be stated concisely:

A plane of current flow between electrodes can be replaced by an insulator.

An equipotential plane can be replaced by an electrode at a fixed potential.

In two dimensions we can consider lines rather than planes to describe surfaces. The first principle is a restatement of the fact that current flows in the same direction as the potential gradient; in a direction perpendicular to current flow, the potential gradient is zero and can be replaced by an insulator. The second principle implies that an equipotential line is equivalent to an electrode at the same potential. Once we make the replacement, we can disregard the portion of the domain on one side of the insulator or electrode.

As an illustration of these principles consider the parallel wire electrodes shown in Fig. 7.7. It is clear from the symmetry that the current must flow directly along the x-axis; therefore, we can replace the x-axis with an insulator. All points equidistant from the centers of the wires must also be equipotential lines. This arrangement implies that the y-axis is an equipotential line, and its potential is the average of the two electrode potentials. The replacement of the x-axis with an insulator is valid for all primary and secondary cases. The replacement of the y-axis with an equipotential is only valid for cases where we maintain symmetry of the field. This symmetry is maintained for a primary current distribution and secondary distributions where the polarization expressions are the same at both electrodes. For a tertiary current distribution, field symmetry would not generally be preserved, and this sectioning would not be valid. By performing these sectioning operations, we have reduced the domain to one-quarter of the original domain. This reduction would decrease the computing effort considerably in a numerical simulation.

A separate issue, not directly related to the sectioning principles, is the simulation of an infinite domain. In this same geometry, we need to establish boundary conditions perpendicular to the axes away from the electrode. Sufficiently far from the electrode, the current flow is negligible, and insulators can be placed perpendicular to the axes. What constitutes a sufficient distance is not immediately apparent. If a large distance is chosen (10 d from the origin), there is little current flow, but the computing effort is large. On the other hand if the insulators are placed close to the electrode (2 d from the origin), the insulators will restrict the current flow, and the calculated current will be less than that in the cell. A pragmatic technique for determining proper placement of the insulators is to run several simulations with the distance of insulator placement as the only variable. When the change is within a specified tolerance, the solution can be accepted.

7.4.3 Numerical Methods

More complex models of electrochemical systems can be treated with general numerical techniques. Effects such as temperature variations, gas evolution, concentration overpotential, and irregular electrode arrangements can all be simulated using numerical methods. A number of general methods are currently available including finite-difference [10], finite element [11], and boundary element [12]. Each method has advantages in terms of ease of implementation, software characteristics (availability, cost, support), hardware requirements, generality, and efficiency.

Commercial software packages are available for all of the common methods, and a choice of package is based on the considerations just mentioned. Alternatively, we could develop computer code for a specific series of simulations. The original simulations of electrochemical cells were based on the finite-difference method [14], and we will illustrate numerical solution techniques with this method.

The basic idea of most numerical techniques is to divide the domain of interest into smaller subdomains. It is invariably easier to find a function that will describe the behavior of a pertinent variable (e.g., potential) over a small domain than it is to find one that describes behavior over the entire domain. Often approximating functions are simple linear or parabolic functions. In some sense continuity must then be maintained among the approximating functions. Because the functionality of the variables is not generally known at the outset, an estimate of the approximating functions—often rather crude—needs to be established. If all goes well, the values of the approximating functions are improved as iteration proceeds. When a specified error criterion is met the solution is said to be converged.

Because the solution is not known *a priori*, an estimate of the error cannot generally be rigorously calculated. Several standard techniques are available for making error estimates. The local error in the approximating functions decreases as the size of the subdomains is reduced; therefore, by running the simulation a second time with smaller subdomains and comparing the values of selected variables, we can determine the magnitude of the error on refinement. If the change is less than some specified criterion, the values of the variables are accepted.

To illustrate the use of numerical methods, we will perform several one- and two-dimensional current distribution simulations using the finite-difference method. The basic concept of the finite-difference method is to cast a differential equation into a corresponding difference equation. The first derivative of a continuous function can be approximated by several forms of difference equations. For example, if we consider the variable ϕ as a function of x, we can subdivide the domain into a number of equal intervals of length h (Fig. 7.13), and approximate the first derivative at the point x_0.

$$\frac{d\phi}{dx} \simeq \frac{\phi_1 - \phi_0}{h} \tag{7.34}$$

$$\simeq \frac{\phi_0 - \phi_{-1}}{h} \tag{7.35}$$

$$\simeq \frac{\phi_1 - \phi_{-1}}{2h} \tag{7.36}$$

The above three approximations are referred to as the forward difference, backward difference, and central difference, respectively. In general, each difference equation yields a different value for the approximate value of the derivative. The error in approximating the continuum with a difference formulation (truncation error) generally decreases as the interval h is reduced; in the forward and backward difference approximations, the error decreases in proportion to h, and in the central difference formula it decreases in proportion to h^2. Although more accurate difference approximations have been developed, they are more complex; their use can only be justified in cases where great accuracy is required.

A second derivative can be approximated by taking differences of the first derivatives, or

$$\frac{d^2\phi}{dx^2} \simeq \frac{\Delta\left(\frac{\Delta\phi}{h}\right)}{h} \tag{7.37}$$

$$\simeq \frac{(\phi_1 - \phi_0)}{h^2} - \frac{(\phi_0 - \phi_{-1})}{h^2} \tag{7.38}$$

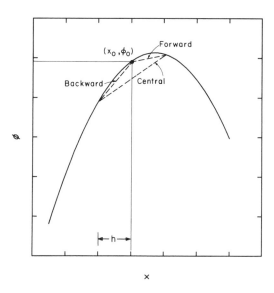

Figure 7.13: Difference approximations for the first derivative at point x_0. The distance between tics on the abscissa is h.

$$\simeq \frac{\phi_1 + \phi_{-1} - 2\phi_0}{h^2} \tag{7.39}$$

For example, if we are treating Laplace's equation in one dimension, Eq. 7.39 can be cast into a particularly simple form:

$$\frac{d^2\phi}{dx^2} = 0 \tag{7.40}$$

$$= \frac{\phi_1 + \phi_{-1} - 2\phi_0}{h^2} \tag{7.41}$$

Because the second derivative is equal to zero, we have

$$\phi_0 = \frac{\phi_1 + \phi_{-1}}{2} \tag{7.42}$$

The error associated with the approximate difference equations cannot be rigorously evaluated, but from a Taylor series analysis, we know that the neglected term is on the order of h^2. As we refine the mesh, we expect to improve the accuracy of the solution. The Taylor series analysis also shows that the error is proportional to the fourth derivative of the function of interest. This relationship implies that mesh refinement will be most effective in problems where potential gradients change most severely.

We already know that the primary potential distribution in one-dimensional cartesian coordinates is a linear function. With linear polarization equations, an explicit expression can be derived for the potential distributions. For most nonlinear polarization equations, the resulting equations can be solved for the potential or current density by a successive approximation technique.

A simple, one-dimensional case can be used to illustrate the principles of an iterative solution technique. For direct comparison, we have analytical solutions. Consider infinite, plane parallel electrodes with linear polarization at the cathode. When the applied potential, conductivity, and electrode separation are specified, we have three unknowns: the potential adjacent to the cathode ϕ_0, the current density, and the overpotential. The three relevant equations are the polarization equation, Ohm's law, and the definition of overpotential:

$$\eta = \beta i \tag{7.43}$$

$$i = \frac{-\kappa(V_a - \phi_0)}{L} \tag{7.44}$$

$$\eta = V_c - \phi_0 \tag{7.45}$$

V_a and V_c are the anode and cathode potentials and L is the electrode separation. Note that ϕ_0 here represents the potential immediately adjacent to the electrode; it should not be confused with the potential at the central node in the finite-difference equations. The current density from these equations is

$$i = \frac{\kappa(V_a - V_c)/L}{1 + \kappa\beta/L} \tag{7.46}$$

and the potential adjacent to the cathode is

$$\phi_0 = \frac{(V_c + \kappa\beta V_a/L)}{1 + \kappa\beta/L} \tag{7.47}$$

A computer program, written in Fortran, which solves this problem by the finite-difference method appears in Appendix F. When we set β to zero ($Wa=0$), the primary current density is obtained. The primary case is unconditionally convergent; we are always assured of obtaining a solution from a properly designed algorithm. As the Wagner number is increased, the number of iterations generally increases (Fig. 7.14). With the simple algorithm used here, the potential in the

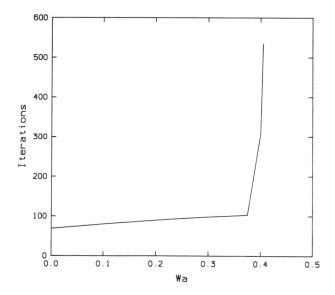

Figure 7.14: Number of iterations required for convergence as a function of the Wagner number. For this algorithm convergence cannot be achieved for Wagner numbers much greater than 0.4.

domain becomes less than the cathode potential at an early iteration for Wagner numbers much greater than 0.4, and the solution becomes unstable. In contrast to the primary current density case, convergence is not guaranteed; however, several techniques are available for increasing the range of convergence. By weighting the potential at a particular iteration with the potential from the previous iteration, wide swings in the potential can be damped. An algorithm to change the weighting factor as a function of the Wagner number and relative changes in the variables has been developed [13].

In this example, potential variation is a linear function. Because the truncation error is proportional to the fourth derivative of the

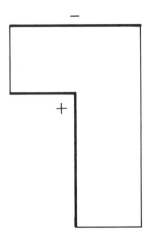

Figure 7.15: Schematic of an L-shaped cell. The heavy lines represent electrodes and the light lines represent insulators.

function, mesh refinement has no significant effect on solution accuracy. This case is, of course, special in that numerical solutions are not generally applied to cells with well-behaved potential functions.

The numerical solution of a two-dimensional secondary current distribution problem is illustrated for an L-shaped cell (Fig. 7.15). This problem was originally treated by Klingert et al. [14]. The geometry can be considered to be a portion of a larger electrode array (Fig. 7.16) that has been reduced by sectioning. In this example we consider only polarization at the right-angled electrode. The program, appearing in Appendix G, is designed to treat cells where the following can be varied: cell dimensions, electrolyte conductivity, and Tafel polarization parameters.

A vector plot of the primary current density distribution on the electrodes appears in Fig. 7.17. Vectors pointing toward the electrode indicate cathodic current densities and those pointing outward indicate anodic current densities. The size of a vector is proportional to the normal component of the current density. For comparison, a secondary current distribution with $Wa = 2.5$ is shown in Fig. 7.18. Both plots are normalized so that the largest vectors are the same length. For the secondary distribution the current density distribution is more

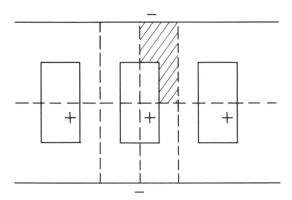

Figure 7.16: Array of electrodes that can be reduced by sectioning to the L-shaped cell.

uniform. This uniformity is especially evident in the recess area on the vertical portion of the electrode. Although the current density is infinite at the corner, the finite-difference approximation indicates a finite current density at that point. Because the potential gradients are changing most rapidly in the corner area, the truncation error is largest there as well. We assume that a more accurate value would be approached with a finer mesh. As in the one-dimensional case, the number of iterations generally increases with the Wagner number. Several small improvements in the program allow us to simulate the cell with Wagner numbers that are greater than in the one-dimensional case.

When electrodes are relatively inert or if electrochemical reactions are slow, the assumption of invariant geometry is acceptable. In some processes the primary objective is to change the electrode geometry. Examples include electroforming and electrochemical machining. In these processes the rate of deposition or dissolution occurs on a relatively long time scale. For example, the deposition rate of copper is on the order of 1 mm/h at a current density of 1 A/cm^2 with 100% current efficiency. The rate of deposition is proportional to the local current density; therefore, we can predict the geometric changes in the electrode shape by moving the electrode boundary in proportion to the local current density. With the new geometry we can calculate a

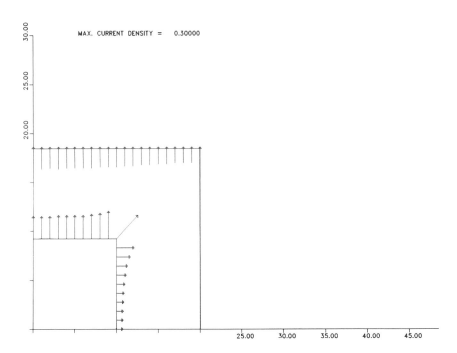

Figure 7.17: Primary current distribution on an L-shaped cell. The magnitude of the local current density is proportional to the size of the vector.

Figure 7.18: Secondary current distribution on an L-shaped cell.

210

new current distribution and move the new boundary in proportion to the new, local current density. The process can be continued through a number of time-steps to simulate the electrode contour change.

As an example of a moving boundary problem, consider the L-shaped electrode, which is similar to the one treated previously. To avoid the difficulty with the sharp corner, we used a slightly rounded geometry. As the simulated deposition process proceeds (Fig. 7.19), the geometry changes, which, in turn, alters the current distribution. By taking relatively small time-steps, we performed a more accurate simulation. The middle profile in the figure represents an intermediate time-step resulting from five successive individual time-steps.

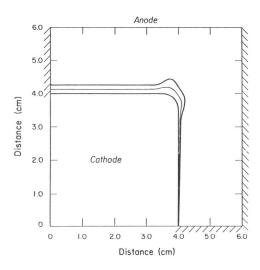

Figure 7.19: Simulated deposition on an L-shaped electrode with a primary current distribution. The two successive profiles represent several time-steps in the simulation.

7.5 Gas-evolving Electrodes

In many important electrochemical processes, gases are evolved at the electrode surface. In the chlor-alkali process, both chlorine and hydrogen are evolved, and the production of aluminum involves the evolution of carbon dioxide at the anode. Metal deposition, electrowinning, and battery charging often involve gas evolution as a side reaction.

There are three primary effects of gas evolution that influence the performance of a cell: (1) gas bubbles reduce electrolyte conductivity and increase ohmic resistance; (2) bubbles that adhere to an electrode block the surface and reduce the area available for reaction; and, (3) rising gas bubbles increase the convection and local heat and mass transfer. The first two effects tend to reduce cell performance, but the third effect tends to increase performance. The relative importance of individual effects depends strongly on current density, cell geometry, hydrodynamics, and the degree of mass transport limitation [15].

Gas bubbles are generally modeled as a suspension of insulating spheres. Theoretical and empirical expressions have been developed for calculating the increased resistance owing to bubbles. One of the early theoretical models was made by Maxwell. He derived a general expression relating the heterogeneous conductivity of a random, dilute suspension of uniform spheres to the void fraction occupied by the spheres.

$$K_m = \frac{1 - f}{1 + f/2} \qquad (7.48)$$

K_m is the ratio of conductivity with bubbles to that of the electrolyte without bubbles present and f is the void fraction. Comparison with experiment revealed that Maxwell's expression gave high accuracy to void fractions less than 0.1. Rather surprisingly, this simple equation does not deviate from experiment by more than about 10% over a range of void fractions up to 80%. It gives this same degree of accuracy even when spheres of different sizes are considered; however, the random dispersion constraint must be maintained. For example, it is clear that a thin, continuous sheet of bubbles could occupy a small void fraction and yet effectively block all current.

Although there are numerous variations on Maxwell's equation, two others can be derived from reasonable models of bubble distributions. The Bruggeman equation was derived by assuming a continuous range of bubble sizes.

$$K_m = (1 - f)^{3/2} \tag{7.49}$$

A third approach was considered by Meredith and Tobias [16]. They recognized that the Bruggeman equation tended to give conductivity values that were slightly lower than experiment in the high void-fraction range. By considering only two sizes of bubbles, they derived the following expression for conductivity:

$$K_m = 8\frac{(1 - f)(2 - f)}{(4 + f)(4 - f)} \tag{7.50}$$

A plot of the three equations in Fig. 7.20 reveals that there is little difference in the conductivity ratio except in higher concentration regions. Most relevant experimental measurements fall between the Maxwell and Bruggeman limits.

The increased mass transport due to gas evolution has been correlated in a number of empirical expressions. Because gas evolution involves strong, local convection, the effects of diffusion and the macroscopic length-scale are of lesser importance; consequently, many of the correlations are presented in terms of the mass transfer coefficient k_m. By contrast, correlations in Ch. 6 are presented in terms of a Sherwood number. If we are to consider mass transfer at a vertical electrode, a correlation of the following form has been proposed [17]:

$$k_m = \left(\frac{\dot{V}}{A}\right)^p \tag{7.51}$$

where $k_m = i_l/(nFc_\infty)$, $\dot{V}$ is the volumetric production rate of gas (cm^3/s), and A is the electrode area (cm^2). The exponent p varies from 0.5 to 0.53 as the current density varies from 5 mA/cm^2 to 30 mA/cm^2.

7.6 Porous Electrodes

We have previously discussed several techniques for increasing the rate of an electrode reaction. Depending on the rate-limiting factors,

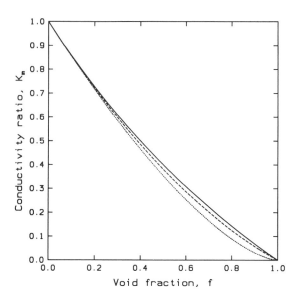

Figure 7.20: Conductivity ratio as a function of void fraction for a random dispersion of spheres from the Maxwell model (solid line), Meredith and Tobias model (dashed line), and Bruggeman model (dotted line).

increased electrolyte agitation, increased temperature, and improved catalysis are standard methods for increasing reaction rates. Another method for increasing the rate per unit electrode area is to use a porous electrode. By using an electrode with a large surface area per unit volume, we can increase the area on which the reaction occurs. Surface areas of 10,000 cm^2 for each cubic centimeter of electrode volume are attainable. With a porous configuration we are able to reduce the electrode volume for a specified total current. Porous electrodes are commonly employed in fuel cells for gaseous reactants. Removal of dilute ionic contaminants from an aqueous solution is also possible with porous electrodes.

The general features of a porous electrode are illustrated in Fig. 7.21. Electrolyte permeates the porous matrix structure. The matrix is electronically conductive, and electrode reactions proceed within the structure. Because of ohmic drop in the electrolyte solution and

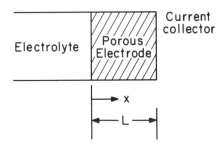

Figure 7.21: Schematic of a porous electrode. All electrochemical reaction occurs within the porous structure.

because reactants must diffuse through the pores, there is a distribution of reaction rates in the porous matrix. The planar metal plate serves as a current collector.

Because we do not know the detailed structure of the pores, we will consider the porous matrix from a macroscopic point of view. Calculations are based on average values of the parameters. This type of macrohomogeneous model was developed by Newman and Tobias [18]. In this model the fundamental transport and kinetic equations have the same basis as those used for planar electrodes, but they are cast in a different form to allow for averaging over the matrix structure.

For models of planar electrodes, we ignored activity in the metal phase. Because we have a long contact path with the electrolyte, we need to consider both the current and potential in the matrix phase. We denote matrix phase variables with the subscript 1 and electrolyte phase variables with the subscript 2. In the following analysis, current densities and fluxes are based on projected area of the electrode. The simple, one-dimensional model with the origin of coordinate system at the metal/solution interface is shown in Fig. 7.21.

In the solution phase we have the following equations developed previously for the flux and current density:

$$N_i = -z_i u_i F c_i \nabla \phi_2 - D_i \nabla c_i + c_i v \qquad (7.52)$$

$$i_2 = F \sum_i z_i N_i \qquad (7.53)$$

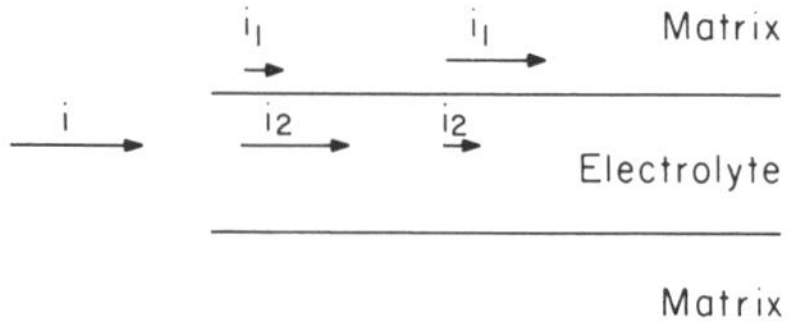

Figure 7.22: Schematic of a single, straight pore. At any vertical plane the total current, represented by the sum of the magnitude of the vectors, is constant.

Because of our perspective, the material balance in the solution phase takes on the slightly different form of

$$\frac{\partial c_i}{\partial t} = -\nabla N_i + \frac{s_i}{nF} \nabla \cdot i_1 \qquad (7.54)$$

The last term arises because we are considering the electrolyte-containing matrix phase as a single entity. Changes in species concentration result from changes in current density in the matrix phase. Changes in current density in the matrix phase as a function of distance can only be due to current leaving or entering the solution phase; current can only enter (or leave) due to reaction, which results in concentration changes. This process is illustrated in Fig. 7.22. Consideration of a single, straight pore reveals that the total current density entering the matrix phase must be equal to the sum of the matrix-phase and electrolyte-phase current densities across any vertical plane at steady-state.

$$i = i_1 + i_2 \qquad (7.55)$$

Differentiating we obtain

$$\nabla \cdot i = \nabla \cdot i_1 + \nabla \cdot i_2 = 0 \qquad (7.56)$$

Because the total current density is constant the derivatives must sum to zero. This expression is a statement of conservation of charge.

The rate at which current enters the solution (di_2/dx in our one-dimensional model) is proportional to the reaction rate on a volumetric basis, or

$$\nabla \cdot i_2 = a\, i_0 \left\{ \exp\left[\frac{\alpha_a F}{RT}(\phi_1 - \phi_2) \right] \right.$$
$$\left. - \exp\left[-\frac{\alpha_c F}{RT}(\phi_1 - \phi_2) \right] \right\} \tag{7.57}$$

where a is the specific interfacial area per unit volume (in units of cm^2/cm^3 or cm^{-1}). In the matrix phase Ohm's law applies

$$i_1 = -\sigma \nabla \phi_1 \tag{7.58}$$

where σ is the matrix phase conductivity. In the solution phase electroneutrality applies

$$\sum_i z_i c_i = 0 \tag{7.59}$$

With simple boundary conditions and Tafel or linear electrode kinetics, a current distribution in the matrix phase can be determined from analytical solutions to the above equations. If we consider Tafel kinetics, the governing differential equation in one dimension follows directly from Eq. 7.57

$$\frac{di_2}{dx} = a\, i_0 \exp\left[\frac{1}{B_a}(\phi_1 - \phi_2) \right] \tag{7.60}$$

where $B_a = RT/\alpha_a F$. One set of boundary conditions is

$$x = 0 \qquad i = i_2, \ i_1 = 0, \ \phi_2 = 0 \tag{7.61}$$
$$x = L \qquad i_2 = 0 \tag{7.62}$$

The conditions at $x = 0$ imply that all of the current is in the electrolyte at the electrode/electrolyte interface. The potential in the electrolyte is defined as zero at that point. At $x = L$ all of the current is assumed to be in the matrix phase, and no current enters the current collector from the solution.

To eliminate the potential we can differentiate Eq. 7.60 with respect to x.

$$\frac{d^2 i_2}{dx^2} = a\, i_0 \exp\left[\frac{1}{B_a}(\phi_1 - \phi_2) \right] \frac{1}{B_a}\left(\frac{d\phi_1}{dx} - \frac{d\phi_2}{dx} \right) \tag{7.63}$$

The term with the exponential is the rate and it is equal to di_2/dx. The derivatives in terms of the potential can each be expressed in terms of the current density according to Ohm's law.

$$\frac{d^2 i_2}{dx^2} = \frac{di_2}{dx} \frac{1}{B_a} \left[-\frac{i_1}{\sigma} + \frac{i_2}{\kappa} \right] \tag{7.64}$$

This equation can also be expressed in terms of i_1. The first derivative of i_1 is the negative of the first derivative of i_2, and $i_2 = i - i_1$.

$$\frac{d^2 i_1}{dx^2} = \frac{di_1}{dx} \frac{1}{B_a} \left[\frac{i}{\kappa} - i_1 \left(\frac{1}{\sigma} + \frac{1}{\kappa} \right) \right] \tag{7.65}$$

This equation can be cast in a dimensionless form with the following definitions:

$$x^* = \frac{x}{L} \tag{7.66}$$

$$i_1^* = \frac{i_1}{i} \tag{7.67}$$

$$\delta = \frac{Li}{B_a} \left(\frac{1}{\kappa} + \frac{1}{\sigma} \right) \tag{7.68}$$

$$\epsilon = \frac{Li}{\kappa B_a} \tag{7.69}$$

Note that ϵ and δ are inverse Wagner numbers, each based on a different weighting of the matrix and solution conductivities. Substitution of these definitions into Eq. 7.65 yields

$$\frac{d^2 i_1^*}{dx^{*2}} = \frac{di_1^*}{dx^*} (\delta i_1^* - \epsilon) \tag{7.70}$$

In terms of the dimensionless variables, the boundary conditions become

$$x^* = 0 \qquad i^* = 0 \tag{7.71}$$

$$x^* = 1 \qquad i^* = 1 \tag{7.72}$$

Eq. 7.70 can be solved by reduction of order with the following result:

$$i_1^* = \frac{2\theta}{\delta} \tan(\theta x^* - \psi) + \frac{\epsilon}{\delta} \tag{7.73}$$

where the integration constants θ and ψ are defined by

$$\tan \theta = \frac{2\delta\theta}{4\theta^2 - \epsilon(\delta - \epsilon)} \tag{7.74}$$

$$\tan \psi = \frac{\epsilon}{2\theta} \tag{7.75}$$

Because the integration constants cannot be determined explicitly, their values must be determined by successive approximation after applying the boundary conditions. The constant θ lies between 0 and π. The dimensionless reaction rate is then determined by differentiation.

$$\frac{di_1^*}{dx^*} = \frac{2\theta^2}{\delta} \sec^2(\theta x^* - \psi) \tag{7.76}$$

The ratio of electrode conductivity to electrolyte conductivity can vary. As an example we can take the conductivities to be equal. A plot of the dimensionless reaction rate as a function of dimensionless distance appears in Fig. 7.23. At high values of δ (low Wagner numbers) the conductivities are relatively low, and much of the current enters the matrix phase near the electrode/electrolyte interface. Current remaining in the solution phase near the current collector must enter the matrix phase, as specified by the boundary condition. Consequently, the current density tends to be high at both interfaces. As δ is reduced, the current distribution becomes more uniform.

An explicit solution to the current distribution problem can be obtained if a linear polarization expression is used. The governing differential equation then becomes

$$\frac{di_2}{dx} = \frac{a}{\beta}(\phi_1 - \phi_2) \tag{7.77}$$

where

$$\beta = i_0 \frac{RT}{(\alpha_a + \alpha_c)F} \tag{7.78}$$

With the same boundary conditions as those used for the Tafel case, the dimensionless current density and reaction rates are

$$\frac{i_2}{i} = \frac{\kappa}{\kappa + \sigma} \left[1 + \frac{\sigma/\kappa \ \sinh \nu(1 - x^*) - \sinh \nu x^*}{\sinh \nu} \right] \tag{7.79}$$

$$\frac{di_1^*}{dx^*} = \frac{\nu\kappa}{(\kappa + \sigma)\sinh \nu} \left[\frac{\sigma}{\kappa} \cosh \nu(1 - x^*) + \cosh \nu x^* \right] \tag{7.80}$$

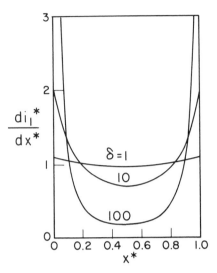

Figure 7.23: Reaction rate distribution in a cell having a matrix-phase conductivity equal to the solution-phase conductivity.

where

$$\nu \;=\; L\left(\frac{a}{\beta}\frac{\kappa+\sigma}{\kappa\sigma}\right)^{1/2} \tag{7.81}$$

Qualitatively, the results are similar at equivalent Wagner numbers. To make comparisons between the linear and Tafel cases, we can choose the same average reaction rate for which

$$\nu \;=\; \sqrt{\delta} \tag{7.82}$$

Use of the linear expression is most appropriate at low current densities, where $|i| << ai_0 L$.

Because diffusive processes are inherently slow, we expect mass transport limitations to affect the performance of a porous electrode. For a dilute solution of the reacting species, reaction should be favored near the electrode/electrolyte interface, where the concentration of reactants is highest. As the reactants become depleted near the current collector, the current density should decrease.

We can account for reactant depletion in a simple way by making the polarization equation concentration-dependent.

$$\frac{di_1}{dx} = ai_0 \frac{c}{c_\infty} \exp\left[-\frac{1}{B_c}(\phi_1 - \phi_2)\right] \tag{7.83}$$

In well-supported electrolyte we can neglect transport by migration, and the flux equation for the reacting species becomes

$$N = -D\frac{dc}{dx} \tag{7.84}$$

The material balance, Eq. 7.54, for the reacting species is

$$D\frac{d^2c}{dx^2} = \frac{1}{nF}\frac{di_1}{dx} \tag{7.85}$$

The boundary conditions on the current and potential are exactly as before (Eqs. 7.61, 7.62), but the following additional conditions must be imposed on the concentration:

$$x = 0 \qquad c = c_\infty \tag{7.86}$$

$$x = L \qquad \frac{dc}{dx} = 0 \tag{7.87}$$

To cast the equations in dimensionless form, we make use of the following additional definitions:

$$\xi = \frac{c}{c_\infty} \tag{7.88}$$

$$\gamma = \frac{iL}{nFDc_\infty} \tag{7.89}$$

Differentiation of Eq. 7.83 yields the same first term as in Eq. 7.70, but in addition there is a second term from the product differentiation of the concentration term, which is also a function of x. In dimensionless terms the differentiated form of Eq. 7.83 is

$$\frac{d^2 i_1^*}{dx^{*\,2}} = \frac{di_1^*}{dx^*}\left(\delta i_1^* - \epsilon + \frac{d\ln\xi}{dx^*}\right) \tag{7.90}$$

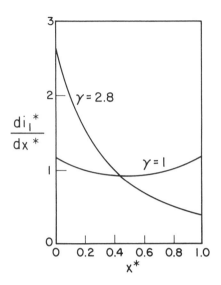

Figure 7.24: Reaction rate distribution in a cell with concentration gradients for $\delta = 2$ and $\epsilon = 0.1$.

Integration of Eq. 7.85 yields

$$\frac{d\xi}{dx^*} = \gamma(i_1^* - 1) \tag{7.91}$$

Eq. 7.90 is nonlinear, and the simultaneous solution of that equation and Eq. 7.91 cannot be accomplished by elementary means; however, standard numerical techniques are available for the simultaneous solution of coupled differential equations. The variable that is proportional to the reaction rate is plotted in Fig. 7.24. The dimensionless variable γ is inversely proportional to the bulk concentration. At larger values of γ (lower bulk concentration), reaction at the electrode/electrolyte interface is favored. At smaller values of γ, the system is not mass transport-limited, and more uniform current distributions are obtained.

7.7 Problems

1. The definition of the Wagner number is given by Eq. 7.4.

a) Derive an explicit expression for the Wagner number when the kinetics are described by the Butler-Volmer equation and $\alpha_a = \alpha_c$. Note that the resulting expression can be expressed in terms of hyperbolic functions as shown in Chap. 5.

b) For $i_0 = 1$ mA/cm^2, the transfer coefficients equal to 0.5, $\kappa = 0.1$ ohm^{-1}-cm^{-1}, and $L = 1$ cm determine the value of the Wagner number when the current density is 10 mA/cm^2.

c) Determine the value of the Wagner number using the Tafel approximation for the same conditions as in Part b.

2. Consider the current distribution on one of the parallel wires shown in Fig. 7.7. Assume that the electrolyte is well-stirred so that concentration gradients are negligible. The electrode kinetics follow the Tafel equation. Determine whether the current distribution becomes more or less uniform when the following changes are made but the total current remains unchanged. Give qualitative reasons for your answers.

a) The temperature is increased.

b) Electrode separation is increased.

c) Supporting electrolyte is added.

3. When individual geometric parameters are varied in multiparameter systems, the primary current distribution changes. In the parallel wire case illustrated in Fig. 7.9, the geometric parameters a and d can both be varied.

a) Use the parameters in the figure as the base case, and calculate the current density distribution when the radius a is decreased by 0.2 cm. Plot i/i_{avg} vs. θ for both cases to compare the normalized current distributions.

b) Increase d to 3 cm, and make the same plot as specified in Part a to compare the current distribution with the base case.

4. Determine the separation of infinite plane parallel electrodes that would have the same average primary current density as the parallel wires shown in Fig. 7.9. Assume that the applied potential and conductivity are unchanged.

5. For the cell shown in Fig. 7.10, determine the primary current distribution on the electrode at $x = 0$.

6. In cylindrical coordinates, Laplace's equation is

$$\nabla^2 \phi = \frac{1}{r} \frac{\partial}{\partial r} \left(r \frac{\partial \phi}{\partial r} \right) + \frac{1}{r^2} \frac{\partial^2 \phi}{\partial \theta^2} + \frac{\partial^2 \phi}{\partial z^2} = 0$$

Consider the case where the field is axisymmetric a) Derive a finite-difference representation for this equation at a radius $r = r_0$. Solve the expression for ϕ_0.
b) Note that for $r_0 = 0$, terms involving $1/r$ become indeterminate. Use l'Hôpital's rule to derive an expression valid along this line.

7. The expressions for determining the current density and potential adjacent to a polarized cathode for infinite, plane, parallel electrodes are given by Eqs. 7.43 to 7.45. The same set of equations can be solved when a Tafel expression, $\eta_s = B \log(i/i_0)$, is used in place of the linear expression Eq. 7.43.
a) Write expressions for the potential adjacent to the cathode and for the current density. Because the Tafel equation is nonlinear, explicit expressions cannot be derived.
b) Write a computer program to determine i and ϕ_0 for $B = 0.1, 1.$, and 10 V. Let $\kappa = 0.1$ ohm^{-1}-cm^{-1}, $L = 1$ cm, $i_0 = 0.001$ A/cm^2, and $V_a - V_c = 1$ V.
c) Plot i and ϕ_0 as functions of the Wagner number.
d) Write a finite-difference program to solve the problem described in Part b. Note that convergence may not be obtainable at higher Wagner numbers with a simple routine.

8. The L-shaped cell can be used as a model for determining the depth to which current can be "thrown" into a recess. Assume that the length of the upper horizontal electrode, shown in Fig. 7.17, is 2 cm and the horizontal portion of the square electrode is 1 cm. Start with a vertical portion of the square electrode equal to 1 cm. Increase the length of this electrode in 1 cm increments until the total cell current increases by less than 1%. Use the following parameters: $\kappa = 0.1$ ohm^{-1}-cm^{-1}, $B = 0.01$ V, $i_0 = 0.001$ A/cm^2, and an applied potential of 1 V. Repeat the problem with $B = 0.1$ V. The computer program in Appendix G can be used in this calculation. The total current can be obtained by integrating the current density over the area.

9. Consider a porous electrode where the polarization expression is linear. The conductivities of the matrix and electrolyte phases are equal to 0.1 ohm^{-1}-cm^{-1}. Assume that the interfacial area per unit volume $a = 100$ cm^{-1}. Plot the dimensionless reaction rate vs. distance into a 1 cm porous electrode for $\beta = 20$ and $\beta = 200$ V/(A/cm^2). Compare the results with those in Fig. 7.23. Note the similarity criterion given in Eq. 7.82.

Bibliography

[1] H. S. Carslaw and J. C. Jaeger, *Conduction of Heat in Solids*, (Oxford: Clarendon Press, 1959).

[2] C. Kasper, *Trans. Electrochem. Soc.*, **77**, 353 (1940); **77**, 365 (1940); **78**, (1940); **82**, 153 (1942).

[3] G. A. Prentice and C. W. Tobias, *J. Electrochem. Soc.*, **129**, 72 (1982).

[4] I. S. Grant and W. R. Phillips, *Electromagnetism* (New York: John Wiley, 1975), pp. 91-4.

[5] P. J. Schneider, *Conduction Heat Transfer* (Cambridge: Addison-Wesley, 1955), pp. 121-28.

[6] C. Wagner, *J. Electrochem. Soc.*, **98**, 116 (1951).

[7] P. Moon and D. E. Spencer, *Field Theory Handbook* (Berlin: Springer Verlag, 1961).

[8] P. J. Sides and C. W. Tobias, *J. Electrochem. Soc.*, **127**, 288 (1980).

[9] J. Newman, *J. Electrochem. Soc.*, **113**, 501 (1966).

[10] M. E. Davis, *Numerical Methods and Modeling for Chemical Engineers*, (New York: John Wiley and Sons, 1984).

[11] J. N. Reddy, *An Introduction to the Finite Element Method* (McGraw-Hill Book Co.: New York, 1984).

[12] C. A. Brebbia, *The Boundary Element Method for Engineers* (Pentech Press: London, 1978).

[13] G. A. Prentice and C. W. Tobias, *AIChE. J.*, **28,** 486 (1982).

[14] J. A. Klingert, S. Lynn, and C. W. Tobias, *Electrochim. Acta,* **9,** 297 (1964).

[15] P. J. Sides, *Modern Aspects of Electrochemistry*, **18,** 303 (1986).

[16] R. E. Meredith and C. W. Tobias, *J. Electrochem. Soc.*, **108,** 286 (1961).

[17] J. R. Selman and C. W. Tobias, *Advances in Chemical Engineering,* **10,** 212 (1978).

[18] J. S. Newman and C. W. Tobias, *J. Electrochem. Soc.*, **109,** 1183 (1962).

Chapter 8

Experimental Methods

A large array of experimental techniques has been developed to probe the behavior of electrode reactions. Many of these techniques have been substantially improved through control and detection made possible by advances in microcomputer technology. Using standard commercial equipment packages, we can make measurements of quantities such as kinetic parameters, mass transfer coefficients, and rates of corrosion.

Several devices are now routinely used in engineering measurements and are available in standard configurations from instrument vendors. Such devices include the rotating disk electrode (RDE), the rotating ring-disk electrode (RRDE), the flow channel, and the hanging mercury drop electrode, all of which have been used in electrochemical measurements for decades. Several specialized instrument packages have become available more recently. Although AC impedance techniques have long been applied to electrode reactions, standard instrumentation packages have become available only in recent years. Other techniques such as ellipsometry and photocurrent measurements are more specialized, and the instrumentation must be custom built from components.

In engineering applications steady-state measurements in flowing electrolyte are most common. Although measurements in initially quiescent electrolyte are easy to carry out, interpretation of results is often obscured by poorly characterized natural convection, leading to fluctuating current-potential behavior. One of the objectives of using flowing electrolyte or a rotating electrode is to maintain well-characterized mass transport conditions. The disk, cylinder, and channel systems have been used in electrochemical studies for many years, and their fluid flow patterns have been described mathematically. Potential and current distributions have been determined, and corrections for ohmic drop and non-uniform current distributions can be estimated.

Techniques based on transient responses are also used in electrochemical studies. Application of potential or current pulses, or sinusoidal potential variations have all been used to estimate kinetic or mass transport parameters. The response of an electrode to electromagnetic radiation has been employed in several contexts. Film structure and growth can be measured by monitoring the changes in phase and amplitude of incident polarized light in ellipsometric determinations. Pulsed light can be used to probe the composition of an electrode film. There are numerous transient techniques for specialized measurements. In the discussion below, we will concentrate on steady-state techniques.

8.1 Rotating Electrodes

Rotating electrodes provide a convenient means to simulate flow conditions for electrochemical measurements in a controlled environment. Current and potential distributions as well as conditions for mass transport limitations have been calculated under a wide range of conditions. The rotating disk and rotating cylinder are the most commonly used in experiments, but the rotating hemispherical electrode has also been used.

8.1.1 Rotating Disk Electrode (RDE)

The RDE has been a standard tool in electrochemical work since the 1940s. It offers many advantages: (1) stable, laminar flow over a wide range of operating conditions; (2) well-established current, potential, and fluid-flow characteristics; (3) uniform limiting current density; and, (4) commercial availability of devices. The RDE is frequently

used in current-potential scans to determine kinetic parameters. At
the limiting current density the diffusivity of an ionic species can be
estimated from pertinent equations. Riddiford [1] has reviewed the
use of the RDE.

A schematic of the RDE appears in Fig. 6.4. Flow across the
surface of the disk is maintained by the pumping action provided by
the rotation. In the z-direction, perpendicular to the plane of the
disk, fluid is drawn toward the surface as a result of the rotation.
In the normal direction the velocity is only a function of the normal
distance from the electrode. This velocity distribution implies that
reactants are transported uniformly to the surface with a resulting
uniform limiting current density.

Velocity components near the RDE have been computed from a
solution of the Navier-Stokes and continuity equations in cylindrical
coordinates. Velocity profiles appear in Fig. 8.1 in terms of the following dimensionless variables:

$$\zeta = z\sqrt{\frac{\omega}{\nu}} \tag{8.1}$$

$$v_\theta = r\omega G \tag{8.2}$$

$$v_r = r\omega F \tag{8.3}$$

$$v_z = \sqrt{\nu\omega}\, H \tag{8.4}$$

At the surface of the disk, the radial velocity component (proportional
to F) is zero because of the no-slip condition. High shear near the
surface results in a velocity maximum for the r-component a small
distance from the surface. The axial component (proportional to H)
must also become zero at the disk surface since the fluid motion is
physically stopped by the solid surface. Far from the disk an asymp-
totic value of $H = -0.88$ is reached; thus implying that a column of
fluid is drawn toward the surface at an essentially constant velocity
until it nears the surface. Because of the no-slip condition, the tangential velocity (proportional to G) attains the same velocity as the
disk at the surface. Far from the disk both the radial and tangential
components are damped out and approach zero.

Flow over a smooth disk is stable at a relatively high value of the
Reynolds number, defined as $Re = r^2\omega/\nu$. Laminar flow prevails up
to $Re = 2\times 10^5$. For a typical disk electrode of less than a centimeter,

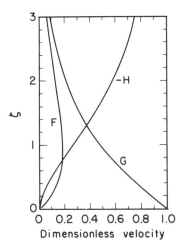

Figure 8.1: Components of velocity on an RDE. The dimensionless distance ζ represents the normal distance from the electrode surface.

laminar flow is maintained at a rotation rate of several thousand rpm. For most experiments, operation in the laminar region is desired.

For operation at the limiting current density in an excess of supporting electrolyte, potential variations can be neglected, and the following steady-state convective-diffusion equation governs the concentration distribution of a reacting species:

$$v \cdot \nabla c = D \nabla^2 c \tag{8.5}$$

Because the normal component of velocity depends only on the normal distance from the electrode, the concentration of reactants transported to the surface is independent of the radial and tangential velocity components. Consequently, the convective-diffusion equation reduces to a one-dimensional problem.

$$v_z \frac{dc}{dz} = D \frac{d^2 c}{dz^2} \tag{8.6}$$

Because the components of fluid velocity can be expressed in series form, an analytical solution to the equation is possible. One boundary condition is that the reactant concentration at the disk surface is zero at the limiting current. The second boundary condition is that the reactant concentration far from the disk is equal to the bulk concentration.

$$c = 0 \quad at \quad z = 0 \tag{8.7}$$

$$c = c_\infty \quad at \quad z \to \infty \tag{8.8}$$

In a typical aqueous solution the kinematic viscosity is on the order of 10^{-2} cm^2/s and the diffusivity is 10^{-5} cm^2/s. The corresponding Schmidt number is on the order of 10^3. A solution to the convective-diffusion equation, valid for large Schmidt numbers, was first presented by Levich and is detailed in his book [2].

$$Sh = 0.62\, Re^{1/2}\, Sc^{1/3} \tag{8.9}$$

This expression is the Levich equation. In terms of the limiting current density, the above equation becomes

$$i_l = 0.62\, nFc_\infty D^{2/3}\nu^{-1/6}\omega^{1/2} \tag{8.10}$$

In a series of experiments, we expect to observe a square root increase in the limiting current density with rotation rate; this observation is one diagnostic to determine whether operation in the mass transfer limited region is occurring.

The Levich equation can also be used to determine the diffusivity from limiting current-rotation rate data. If the electrode reaction is known, all of the other parameters are easily determined or measured. One problem with determining the diffusivity is that the concentration is varying in the boundary layer, and the diffusivity generally decreases with increasing concentration. Because the concentration distribution varies with bulk concentration and flow conditions, the value of the diffusivity also varies with these factors. The value obtained by the rotating disk method represents an integrated average over the concentration range and is referred to as the integral diffusion coefficient.

The RDE is also a useful tool below the limiting current. It is often desirable to minimize the effects of mass transport limitations when kinetic parameters are being studied. The disk allows operation at low fractions of the limiting current density by increasing the rotation rate. The primary current distribution [3] and secondary current distributions [4] over a wide range of Wagner numbers have been determined. The expression for the primary current distribution is

$$\frac{i}{i_{avg}} = \frac{0.5}{\sqrt{1 - r^2/r_0^2}} \tag{8.11}$$

where r_0 is the disk radius. The resistance between a disk and hemispherical counterelectrode at infinity is

$$R = \frac{1}{4\kappa r_0} \tag{8.12}$$

Expressions for the secondary current distributions using linear and Tafel equations for the surface overpotential were derived by Newman [4]. For Tafel kinetics the Wagner number is defined as

$$Wa = \frac{\kappa B}{r_0 \, i_{avg}} \tag{8.13}$$

This definition of the Wagner number is the inverse of the quantity J in Newman's paper. A graphical representation of several current distributions as a function of Wagner number appears in Fig. 8.2. The primary current distribution ($Wa = 0$) is the most non-uniform. At the center of the disk, the current density is half the average current density. Because the electrode is coplanar with the insulating surface, the current density becomes infinite at the edge for the primary current distribution. As the Wagner number is increased, the current density becomes more uniform. The non-uniform current distribution below the limiting current is one of the least desirable features of the RDE.

With a non-uniform current distribution, we generally assume that the average current density adequately approximates the behavior of an electrode with a uniform current density of the same value. Alternatively, experiments can be designed to maximize the Wagner number. If supporting electrolyte does not interfere with the electrode reaction, its concentration can be increased. Reduction of the electrode radius also increases the Wagner number.

Because the potential distribution can also be calculated, some general guidelines to placing the reference electrode can be formulated.

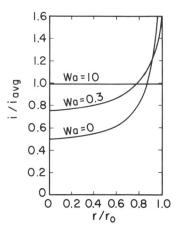

Figure 8.2: Secondary current distribution for Tafel kinetics on an RDE.

For the primary current distribution, Newman [3] showed that the potential gradient is very high near the disk, especially adjacent to the edge of the electrode. In one example, he showed that placement of a reference electrode at a distance equal to several electrode diameters from the electrode along the axis was sufficient to avoid high fields. At a distance equal to five electrode diameters the resistance was within 10% of the resistance to a hemispherical counterelectrode infinitely far away. Newman also demonstrated that placement of the reference electrode in the plane of the disk, five diameters from the center, gave approximately the same result.

8.1.2 Rotating Ring-disk Electrode (RRDE)

One variation on the rotating disk electrode is the rotating ring-disk electrode (RRDE) [5]. It consists of a ring electrode concentric to the disk as shown in Fig. 8.3. With this arrangement the ring and the disk can be held at different potentials. A primary use of this device is to detect reaction products created at the disk. For example, we may have the following reaction scheme:

$$A \ \rightarrow \ B \ \rightarrow \ C \tag{8.14}$$

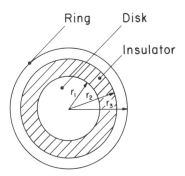

Figure 8.3: Schematic of a rotating ring-disk electrode.

If B is produced at the disk, then the ring electrode can be held at a potential to convert B back to A. By measuring the current at the ring, we can make a quantitative determination of the fraction of A converted to B. For example, a proposed mechanism for the reduction of oxygen was confirmed using the ring-disk electrode. The overall reaction is

$$O_2 + 4H^+ + 4e = 2H_2O \qquad (8.15)$$

Two simpler overall reactions resulting from several intermediate reactions are

$$O_2 + 2H^+ + 2e = H_2O_2 \qquad (8.16)$$
$$H_2O_2 + 2H^+ + 2e = 2H_2O \qquad (8.17)$$

Because hydrogen peroxide is a stable intermediate, it can be detected by oxidizing it on the ring.

Under ideal conditions not all of the B produced at the disk is converted back to A at the ring because not all of the fluid containing B contacts the ring. The ratio of the current on the ring to that on the disk is called the collection efficiency and can be expressed as

$$N_0 = \frac{I_r}{I_d} \qquad (8.18)$$

In this definition we treat all currents as positive quantities. The derivation of a theoretical expression follows from a solution of the convective diffusion equation (Eq. 8.5). It was first derived by Albery and Bruckenstein [6].

$$\begin{aligned} N_0 &= 1 - F(\alpha/\beta) + \beta^{2/3}[1 - F(\alpha)] - \\ &\quad (1 + \alpha + \beta)^{2/3}\{1 - F[(\alpha/\beta)(1 + \alpha + \beta)]\} \end{aligned} \tag{8.19}$$

where

$$\alpha = (r_2/r_1)^3 - 1 \tag{8.20}$$

$$\beta = (r_3/r_1)^3 - (r_2/r_1)^3 \tag{8.21}$$

$$\begin{aligned} F(\theta) &= \frac{3^{1/2}}{4\pi} \ln\left[\frac{(1 + \theta^{1/3})^3}{1 + \theta}\right] + \\ &\quad \frac{3}{2\pi} \arctan\left(\frac{2\theta^{1/3} - 1}{3^{1/2}}\right) + \frac{1}{4} \end{aligned} \tag{8.22}$$

Collection efficiencies are tabulated in the book by Albery and Hitchman [5]. For example, for $r_1 = 0.4769$ cm, $r_2 = 0.4869$ cm, and $r_3 = 0.5221$ cm, the collection efficiency is 0.214. This value implies that under ideal conditions only 21% of the electroactive material produced on the disk can be detected on the ring. Although theoretical calculation of the collection efficiency often agrees within a few percent with experimental results, an experimental check with a redox couple is frequently carried out. The ferri/ferrocyanide couple provides a simple system for calibration of the device.

$$Fe(CN)_6^{4-} = Fe(CN)_6^{3-} + e \tag{8.23}$$

8.1.3   Rotating Cylinder Electrode (RCE)

A rotating cylinder electrode (RCE) apparatus is another convenient tool for use in studying electrode kinetics, ionic mass transfer, and corrosion rates [7]. The commercial apparatus used for the RDE is easily modified to accommodate cylindrical electrodes. Typically, an inner cylinder of 1- to 3-cm diameter with an outer counterelectrode on the order of 10-cm diameter is used for measurements (see Fig. 6.5). We would choose the RCE over the RDE if experiments in the turbulent flow region were desired. With proper design a uniform current distribution can be obtained.

Conventional RCE systems are designed so that the inner cylinder rotates and creates a turbulent flow in the fluid. It is possible to rotate the outer cylinder and maintain greater flow stability, but this design causes operational difficulties. When the inner cylinder is rotated slowly, a laminar flow pattern can be maintained; however, in laminar flow fluid moves in a circular pattern about the axis with no radial component. Thus, there is no convection in the radial direction, and no enhancement of mass transport due to fluid flow. Consequently, an RCE is rarely operated in the laminar flow regime in electrochemical studies.

Several different criteria have been proposed to characterize the flow regime. These include Reynolds numbers with different length dimensions (inner diameter and interelectrode gap) and Taylor numbers. A Reynolds number with the inner diameter as the length-scale provides an effective measure of fluid characteristics in most systems.

$$Re = \frac{d_i v}{\nu} \qquad (8.24)$$

where d_i is the inner cylinder diameter and v is the surface velocity of the inner cylinder. If the rotation rate is expressed in rpm, then the surface velocity is

$$v = \frac{\pi d_i}{60} \text{ rpm} \qquad (8.25)$$

With the Reynolds number defined by Eq. 8.24, laminar flow prevails on smooth cylinders for $Re < 200$. A transition region occurs for Reynolds numbers between 200 and 2000. In this flow regime a series of toroidal cells, known as Taylor vortices, form in the interelectrode gap. Operation in this transitional regime is undesirable in electrochemical studies. For a Reynolds number greater than 2000, fully developed turbulent flow is maintained, and mass transport is significantly enhanced as the rotation rate is increased. These criteria apply to smooth cylinders. For cylinders that become roughened, usually as a result of deposition or dissolution, correlations in terms of friction factors can be used to determine flow regime. Several of these correlations are reviewed by Gabe [7].

The original correlation for determining limiting current density as a function of dimensionless groups was formulated by Eisenberg,

Tobias, and Wilke [8] as

$$Sh = 0.0791\, Re^{0.7}\, Sc^{0.356} \tag{8.26}$$

In terms of the limiting current density this equation becomes

$$i_l = 0.0791\, nFc_\infty v^{0.7} d_i^{-0.3} \nu^{-0.344} D^{0.644} \tag{8.27}$$

With an RDE the limiting current density increases with the square root of rotation rate, whereas for an RCE the limiting current density increases with the 0.7 power of the rotation rate.

Because a potential distribution is only a function of distance, the ohmic drop to a reference electrode is readily calculated. For an RDE potential distribution changes with total current, and ohmic compensation is somewhat more complicated. If the conductivity and distance between the working and reference electrodes are known, the ohmic drop is given by

$$\Delta\phi_{ohm} = \frac{I}{2\pi H \kappa}\, \ln\frac{r}{r_i} \tag{8.28}$$

where r is the radial distance from the inner cylinder.

8.2 Flow Channel

Systems where fluid circulates past stationary electrodes are also used in electrochemical studies. The best-characterized of such systems is the flow channel, where fluid is pumped between plane, parallel electrodes embedded in insulating walls. Fig. 8.4 illustrates this arrangement. Useful operating ranges in both the turbulent and laminar regimes are provided with the flow channel. From an operational viewpoint a channel is more cumbersome; a reservoir, pumps, and multiple reference electrodes are usually required. An advantage is that many industrial processes utilize plane, parallel electrodes, and the channel may provide the most realistic simulation of actual electrode performance.

Fluid flow patterns in the laminar regime are well-characterized. A linear velocity gradient is usually assumed at the electrode surface.

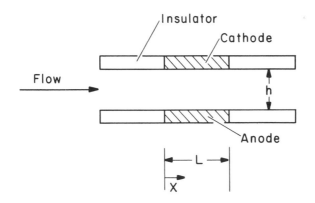

Figure 8.4: Plane parallel electrodes embedded in insulating walls.

For this analysis we can consider the case of a reduction reaction at the upper cathode. Because fresh reactants enter from the left, the current density is highest at the leading edge. The mass transfer boundary layer increases as the fluid becomes depleted, and the current density on the cathode decreases as the fluid proceeds down the channel. The rate of decrease depends on the degree of mass transport control. In some cases, where mass-transport limitations are not severe, the current density near the trailing edge can increase due to edge effects.

The criterion for laminar flow is determined by the value of the Reynolds number.

$$Re = \frac{d_e \, v_{avg}}{\nu} \tag{8.29}$$

where d_e is an equivalent diameter defined by

$$d_e = \frac{4 \, A_{cs}}{P} \tag{8.30}$$

where A_{cs} is the cross-sectional area of the channel and P is the perimeter. For a cell with an interelectrode gap equal to h and width equal to w the equivalent diameter is $2 \, wh/(w + h)$. The transition from laminar to turbulent flow occurs at a Reynolds number of approximately 2000; the value varies somewhat depending on cell dimensions and electrode roughness.

To maintain laminar flow in a cell, an entry region must be included in the system design. Generally, the entry length is approximately 40 equivalent diameters. For a tubular cell the entry length can be predicted from the correlation

$$\frac{L_{ent}}{d} = 0.058\,Re \tag{8.31}$$

where d is the tube diameter. An approximate result can be obtained by substituting d_e for d in this equation.

Limiting current density relations in terms of dimensionless variables have been presented by Parrish and Newman [9]. For electrodes where the separation is much greater than the length, the following relation applies:

$$Sh = 1.85 \left(\frac{Re\,Sc\,d_e}{L}\right)^{1/3} \tag{8.32}$$

where the characteristic dimension for Re and Sh is the equivalent diameter. In terms of the average limiting current density, the above equation becomes

$$i_l = 1.85\,nFDc_\infty \left(\frac{v_{avg}}{d_e DL}\right)^{1/3} \tag{8.33}$$

Because the boundary layer grows down the electrode, the trailing portion is less effective than the leading portion of the electrode; consequently, the average current density varies with the inverse cube root of electrode length. Also, the average limiting current density varies with the cube root of average fluid velocity.

The current distribution on the electrode was also derived by Newman as

$$Sh(x) = 1.23 \left(\frac{Re\,Sc\,d_e}{x}\right)^{1/3} \tag{8.34}$$

In terms of local current density, this equation becomes

$$i_l(x) = 1.23\,nFDc_\infty \left(\frac{v_{avg}}{d_e Dx}\right)^{1/3} \tag{8.35}$$

The current distribution is given by

$$\frac{i}{i_{avg}} = \frac{2}{3} \left(\frac{x}{L}\right)^{-1/3} \tag{8.36}$$

The primary current distribution is the extreme case of no concentration gradients. A comparison of this case with the mass transport-limited case appears in Fig. 8.5. Because the field is symmetric about the midplane of the electrode at $x = L/2$, the primary current distribution is also symmetric about that plane. By contrast, the concentration distribution is asymmetric. The concentration of reactant is highest at the leading edge, and the local current density reflects the concentration variations under mass-transport limitations.

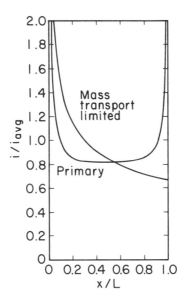

Figure 8.5: Current distributions on the cathode of plane parallel electrodes. Calculations are for the case where $L = 2h$.

8.3 Problems

1. Copper is being deposited on a 1-cm diameter rotating disk electrode from 0.2 M $CuSO_4$ and 0.5 M H_2SO_4 electrolyte at room temperature.

a) Estimate the limiting current density at 500 rpm.

b) Estimate the current density at the center of the disk when the rotation rate is doubled. Assume that the total current is the same as in Part a.

2. The ability to detect intermediates on the ring of a ring-disk electrode depends on the stability of the intermediate. If $r_1 = 0.4$ cm, and $r_2 = 0.5$ cm (Fig. 8.3), estimate the minimum lifetime of an intermediate required for detection. Assume a rotation rate of 500 rpm.

3. Under the conditions specified in Problem 1b, estimate the ohmic drop (one significant figure) between the disk and a reference electrode located 5 cm from the disk on its axis.

4. One source of error in the measurement of ohmic drop results from inaccurate measurement of the distance between the working and reference electrodes. Consider a rotating cylinder electrode system with an inner radius of 1 cm, an outer radius of 5 cm, height of 10 cm, and with electrolyte that is 0.05 M KOH. The total current is 1 A. Estimate the rate of change of potential (mV/mm) with distance for a reference electrode located 1 mm from the inner electrode. Repeat the calculation for a reference electrode 2 cm from the inner electrode.

5. Make a qualitative sketch of the secondary current distribution for Wagner numbers of 0.1, 1, and 10 in a plane parallel electrode geometry. Use Fig. 8.5 as a model.

Bibliography

[1] A. C. Riddiford, *Advances in Electrochemistry and Electrochemical Engineering*, **4**, 47 (1966).

[2] V. G. Levich, *Physicochemical Hydrodynamics* (Prentice Hall: Englewood Cliffs, NJ, 1962) pp. 60-78.

[3] J. Newman, *J. Electrochem. Soc.*, **113**, 501 (1966).

[4] J. Newman, *J. Electrochem. Soc.*, **113**, 1235 (1966).

[5] W. J. Albery, *Ring-disc Electrodes* (Clarendon Press: Oxford, 1971).

[6] W. J. Albery and S. Bruckenstein, *Trans. Faraday Soc.*, **62**, 1920 (1966).

[7] D. R. Gabe, *J. Appl. Electrochem.*, **4**, 91 (1974).

[8] M. Eisenberg, C. W. Tobias, and C. R. Wilke, *J. Electrochem. Soc.*, **101**, 306 (1954).

[9] W. R. Parrish and J. Newman, *J. Electrochem. Soc.*, **116**, 169 (1969).

Chapter 9

Applications

In the first chapter we discussed some of the areas where electro-chemical engineering plays a role: electrolysis, energy conversion and storage, and corrosion. We will now study a few specific applications to gain a greater appreciation of the overall characteristics of electrochemical systems. In addition to the technical factors, we need to consider costs relative to competing technologies. In commercial applications, factors such as energy efficiency, power density, or other technical factors may be required for an energy storage system, but ultimately cost considerations may dictate the use of a particular system. Some electrode combinations are inherently more energetic. This factor is of particular importance in transportation applications, where a low specific energy may rule out a particular electrode couple.

9.1 Energy Storage and Conversion

Common batteries such as the carbon-zinc (Leclanchè) and lead-acid batteries have been used in a number of general applications. These devices, along with fuel cells, convert chemical energy directly to electrical energy. Several classes of energy storage and conversion devices are used in various applications. Batteries are considered to be storage devices in the sense that they only use the chemical reactants originally supplied. Batteries that can be recharged are called secondary batteries as opposed to those that are used only one time (primary batteries). Nickel-cadmium and lead-acid cells are examples

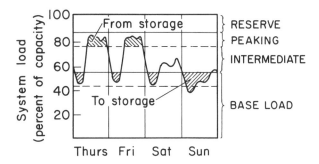

Figure 9.1: Utility load-leveling. Power generated during off-peak hours could be returned to the grid during peak hours.

of secondary systems, while the Leclanchè cell is a common primary battery. With a fuel cell, reactants are continuously supplied from an external source, and the reaction products are generally discarded. Several fuel cell and advanced battery systems are being considered for large-scale application in electricity generation and transportation.

9.1.1 Utility Load-leveling

In an electric utility, advanced batteries are being considered for load-leveling. The concept is to run a conventional power plant at a relatively constant output, storing the off-peak energy in a battery and returning electricity to the grid during peak periods (Fig. 9.1). Typically, batteries would be charged between midnight and 6 A.M., and discharged for about five hours during peak demand. Operation in this base-load mode is most efficient, and the capital cost associated with additional load-following generating equipment is avoided. Batteries are particularly well-suited to this application because they can easily be controlled to follow short-term demand fluctuations. They can be constructed in small modules, allowing the addition of units in increments of a few MW. Because they can be dispersed as modules, less transmission equipment is required. If intermittent sources of electric power such as wind or solar become more prevalent, batteries could be used to balance demand.

A battery system for load-leveling would need to have a low capital cost and long cycle life. To compete with existing technologies, an advanced battery would need to be constructed to last for at least 2500 cycles (10 years) at a cost of approximately \$50/kW-h. A lead-acid battery can be cycled for approximately 500 times when discharged to 80% of its capacity. The cost is on the order of \$100/kW-h. Other systems including $LiAl/FeS_2$, Na/S, and Zn/Cl_2 are being considered for load-leveling as well.

9.1.2   Electric Vehicles

The use of electrochemical devices to power electric vehicles has many attractive features: high efficiency, low environmental impact, and low noise. In addition to the low cost and high cycle life required for load-leveling, there are requirements of high specific energy and specific power. If specific energy is low, then a large fraction of the total vehicle weight is required for the battery. Low specific power implies that the ability to accelerate would be poor. Specific energy is closely related to cell thermodynamics. The theoretical specific energy (TSE) is defined as the maximum energy available from an electrochemical device divided by reactant mass

$$TSE \; = \; \frac{\Delta G_r}{\displaystyle\sum_{reactants} m_i} \tag{9.1}$$

Reactant mass is usually based on reactants carried with the device. Determination of reactant mass can vary with the application. For example, reactant mass of a hydrogen fuel cell in an electric vehicle would be based on the mass of hydrogen as oxygen would be obtained from the air. In a spacecraft oxygen would be carried along and included in a calculation of the reactant mass.

As a general guide, the available specific energy is approximately 20% of the theoretical specific energy. The specific power available from a device is a function of electrode kinetics, cell design, and mass transport characteristics. A plot of specific power vs. specific energy appears in Fig. 9.2. The curves reflect typical performance characteristics, but could be improved through engineering advances. The

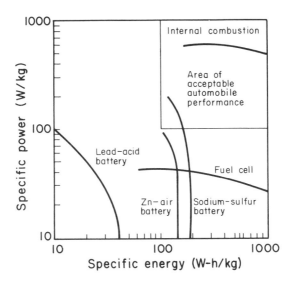

Figure 9.2: Ragone plot. Specific energy generally decreases as the load (power requirement) is increased.

specific power reflects the rate of discharge. It could be increased dramatically by reducing factors such as ohmic losses and overpotential. In principle, specific power is limited by electrode passivation, mass transport-limitations, and ohmic and kinetic losses. There is no well-defined limit to specific power. By contrast, the specific energy has a well-defined upper limit given by the theoretical specific energy.

For acceptable vehicle performance a specific energy of 100 W-h/kg and a specific power of 100 W/kg are required. Lower specific energy and power would be adequate for a vehicle that would only be operated in an urban environment. Low specific energy in a device results, in large part, from the low theoretical specific energies of most aqueous systems. In general, metals that are most active are not stable in aqueous media. Although some nonaqueous electrolytes are used with more active metals such as lithium, the lower conductivity of those media reduces specific power.

From Fig. 9.3 we see that Zn/O_2 is a highly energetic couple with a theoretical specific energy of approximately 1200 W-h/kg. The theoretical specific energy is also high because the oxidant (air) is not

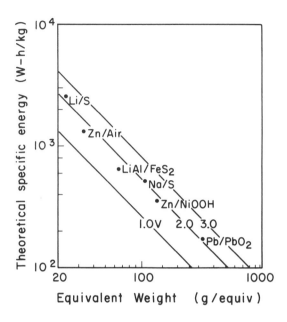

Figure 9.3: Theoretical specific energy for several couples. Low density, active materials are capable of producing more energy per unit weight.

carried on the vehicle. Systems with theoretical specific energies in excess of 1000 W-h/kg are primarily high-temperature, molten salt systems. Many of the systems under development that meet electric vehicle requirements are high-temperature, molten salt systems such as Na/S and LiAl/FeS$_2$. High-temperature operation also promotes electrode kinetics, resulting in increased specific power.

The design parameters for a small urban vehicle with minimum performance have been determined [1]. At the axle the energy required is about 0.12 kW-h/T-km and the average power required is 25 kW/T. We can estimate the battery requirements for an electric vehicle based on the following requirements: a 1 T (1000 kg) vehicle, battery weight equal to 30% of vehicle weight, 80 km range, 80%

battery-to-axle efficiency. The energy required is

$$\frac{(0.12kW - h/T - km)(1T)(80km)}{(0.3Tbattery)(0.8)} = 40W - h/kg \qquad (9.2)$$

The specific power is

$$\frac{(25kW/T)(1T)}{(0.3T)} = 83W/kg \qquad (9.3)$$

A projected cost for a vehicle battery that lasts 1000 cycles (3 years) is approximately $75/kW-h. In our case the energy requirement is 15 kW-h, and the battery cost is about $1000.

We can make a crude comparison of energy costs between an electric vehicle and a conventional internal combustion engine. If we choose as a basis a vehicle that travels 30,000 km/y, the energy used is 4500 kW-h. For electricity at $.08/kW-h, the energy cost is $360, and about one-third of the battery has been used; the total energy cost is about $700 or $.02/km. The gasoline consumption for a similar conventional vehicle would be 80 km/gal. At $1/gal for gasoline, the energy cost is about $ .01/km.

The above calculation neglects many factors that are difficult to quantify. A large credit should be taken for the pollution reduction in urban areas when using an electric vehicle. The additional pollutants produced at a central generating facility are much easier to control than pollutants from a large number of individual cars. Recharging of batteries during off-peak periods would be a form of load-leveling that would contribute to the efficiency of power plants. On the other hand, the range and performance of an electric vehicle would be inferior to the conventional auto. The 25 kW power of the electric vehicle corresponds to about 33 hp. The energy cost comparisons do not take into account differences in original vehicle cost. For the vehicle specified above, fixed costs would dominate the overall driving cost. The primary factor that limits acceptance of electric vehicles is the restricted range and the inability to recharge rapidly. From a technical perspective the range could be increased by developing a system with a significantly higher specific energy. Although molten salt systems have this capability, the problems associated with thermal management and corrosion have not been overcome.

9.1.3   Fuel Cells

Fuel cells are being developed for electric power generation and for transportation applications. The primary reasons that fuel cells have not seen widespread use in these applications are similar to those for advanced secondary batteries: relatively high cost and low specific power. From Fig. 9.2 we see that specific energy is very high, and because of this factor, fuel cells have been chosen as the primary energy source on spaceflights.

Pilot-scale fuel cells have been successfully tested in demonstration projects. The most intensive development has been focused on the phosphoric acid system operating at 200°C with hydrogen fuel reformed from naphtha. The inherent efficiency of the fuel cell is higher because Carnot losses are eliminated. Conversely, low-temperature devices are subject to severe kinetic limitations, especially at an oxygen electrode. Although the energy efficiency of a single cell increases with decreasing load, the overall efficiency of a fuel cell-based power plant is relatively constant over a fairly wide operating range (Fig. 9.4).

Most systems under development rely on hydrogen as the fuel and oxygen from the air as the oxidant. Although direct use of a hydrocarbon or methanol would be highly desirable, the kinetics of these reactions are too sluggish to permit operation at a reasonable rate. In present phosphoric acid systems the oxygen electrode is responsible for much of the voltage loss. The current density-potential plot (Fig. 9.5) for a phosphoric acid fuel cell shows that the overpotential associated with oxygen reduction is greater than the total of all other sources of irreversibility. At a typical current density of 200 mA/cm^2, terminal voltage is about 0.65 V or roughly half the reversible potential. Terminal voltage is a key factor in predicting efficiency. Overall efficiency ϵ_e from fuel to electrical power at the operating current density can be crudely estimated from the expression

$$\epsilon_e = 59\, V_t \qquad (9.4)$$

where V_t is the terminal voltage.

To overcome kinetic limitations, some effort is directed toward the development of high-temperature, molten salt fuel cells. A system

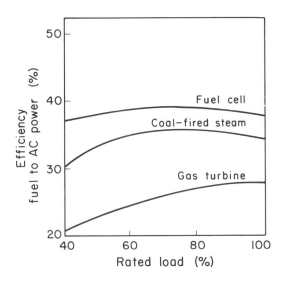

Figure 9.4: Power plant efficiency tends to be fairly constant for a fuel cell over a wide range of loads.

with a mixture of molten K_2CO_3/Li_2CO_3 operating at 650^0C is significantly more efficient than a low-temperature fuel cell. Because of rapid kinetics, expensive catalysts can be avoided and high hydrogen purity, required to avoid catalyst poisoning, is also unnecessary. Although these systems use hydrogen as the fuel, the electrode reactions involve CO_2.

$$H_2 + CO_3^{2-} = CO_2 + H_2O + 2e \qquad (9.5)$$

$$CO_2 + \frac{1}{2}O_2 + 2e = CO_3^{2-} \qquad (9.6)$$

A severe drawback of these systems is the corrosion of fuel cell components in these aggressive environments.

We can make some crude estimates of costs and compare them with the costs of running a conventional electric power plant. The capital cost for a fossil fuel-generating facility is approximately \$3000/installed kW. Because fuel cell systems require pre-processing of their fuel, the fuel cell itself represents only about one-quarter of the total capital cost. On this basis the fuel cell cost should be less than

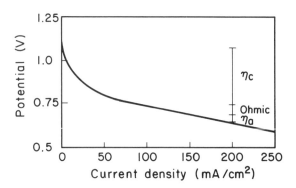

Figure 9.5: Voltage losses in a phosphoric acid fuel cell as a function of current density.

$800/kW. Typically, terminal voltage is 0.5 V and current density is 200 mA/cm^2, which implies that the power density is about 100 mW/cm^2. Each of the two electrodes would need to be 10,000 cm^2 for each kW. To meet the $800/kW criterion, the fuel cell system, based on electrode area, could cost no more than $.04/cm^2 or about $40/ft^2. This cost goal presents a difficult challenge. The current generation of phosphoric acid fuel cells requires platinum loading of 0.1 to 1 mg/cm^2; this cost alone represents about $.01/cm^2. A second practical issue that needs to be addressed is whether there is enough of a particular material to meet anticipated demand for a specified system. With platinum as a catalyst, the number of power plants that could be constructed with estimated reserves would be sufficient only for a small fraction of the total electrical demand.

9.1.4   Thermally Regenerative Electrochemical System (TRES)

Most forms of large-scale electrical energy production rely on the conversion of thermal energy into mechanical and subsequently, electrical energy. A variation on this concept is to run two electrochemical devices (fuel cells) at different temperatures, as shown in Fig. 9.6. At the high-temperature cell the reaction is run in the forward di-

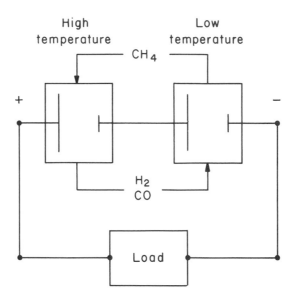

Figure 9.6: Schematic of a TRES. Products generated from the high-temperature fuel cell become the reactants for the low-temperature fuel cell.

rection, and at the low-temperature cell the same reaction is run in the reverse direction. This system depends on identifying a reaction where the reversible potential varies with temperature. Such a reaction would have a large entropy change of reaction. With this reaction we could use the products of the high-temperature cell as reactants for the low-temperature cell. To complete the loop, the products from the low-temperature cell would be heated and sent back to the high-temperature cell. This process is called a Thermally Regenerative Electrochemical System (TRES). Although a commercial system has not been developed, the TRES concept has many of the advantages of the fuel cell and could be considered if a suitable reaction were identified.

The reaction of steam with methane was proposed by Rapp [2] as a candidate reaction for use in the TRES.

$$CH_4 + H_2O \rightarrow 3H_2 + CO \tag{9.7}$$

This reaction can be run in an electrochemical cell at high temperature (1200 K) with the following reactions at the anode and cathode, respectively:

$$CH_4 + 3\,O^{2-} \rightarrow CO + 2\,H_2O + 6e \tag{9.8}$$

$$3\,H_2O + 6e \rightarrow 3\,H_2 + 3\,O^{2-} \tag{9.9}$$

At low temperature (400 K), Reaction 9.7 can be run in reverse with the following electrode reactions at the anode and cathode, respectively:

$$3\,H_2 \rightarrow 6\,H^+ + 6e \tag{9.10}$$

$$CO + 6\,H^+ + 6e \rightarrow CH_4 + H_2O \tag{9.11}$$

The entropy change of reaction is sufficiently large so that the forward reaction is thermodynamically favored at the high-temperature cell and the reverse reaction is favored at the low-temperature cell. At 1200 K, $\Delta G = $ -76 kJ for Reaction 9.7, while at 400 K, $\Delta G = $ -126 kJ for the reverse reaction. The enthalpy changes of reaction are +227 kJ for the forward reaction at 1200 K and -227 kJ for the reverse reaction at 400 K.

The reversible potential for the fuel cell at 1200 K is

$$
\begin{aligned}
E &= -\frac{\Delta G}{nF} \\
&= -\frac{-76,000}{(6)(96,500)} \\
&= 0.13 \text{ V} \tag{9.12}
\end{aligned}
$$

At 400 K the reversible potential is 0.22 V. A system operating reversibly would maintain a potential that is the sum of the potential of the two cells, or 0.35 V.

Because this process relies on the absorption of heat at the high temperature (T_h) and rejection of heat at the low temperature (T_l), it is subject to Carnot limitations. The maximum overall efficiency for this process is

$$\epsilon_e = \frac{T_h - T_l}{T_h} \tag{9.13}$$

$$\epsilon_e = \frac{1200 \text{ K} - 400 \text{ K}}{1200 \text{ K}}$$
$$= 0.67 \tag{9.14}$$

Although the TRES concept is attractive in its simplicity, it is difficult to identify a reaction with the desirable characteristics. The methane reaction has a fairly high-temperature coefficient of change for the reversible potential, but the reversible potential of only 0.35 V is insufficient for a commercial process. Because of kinetic limitations at the low-temperature electrode and other irreversibilities, we can only expect an operating potential of about half the reversible potential.

9.2   Electrolytic Processes

The major electrolytic processes are the reduction of aluminum and the synthesis of chlorine and caustic from brine. There is currently one major electroörganic process, the hydrodimerization of acrylonitrile to form adiponitrile, expressed chemically as

$$2\,CH_2CHCN + 2\,H_2O + 2e \rightarrow CN(CH_2)_4CN + 2\,OH^- \tag{9.15}$$

Adiponitrile is an intermediate in the synthesis of nylon. About 10^6 ton/y are produced by this route in the US. No other electroörganic syntheses approach this output.

Although thousands of electroörganic reactions have been cataloged [3], few have been developed for commercial use. The potential advantages of an electrochemical route include high efficiency, good process control, mild operating conditions, and minimal environmental effects. Adoption of electrochemical syntheses has been relatively slow, in part, because of the general lack of experience in process design. Practical design considerations such as cell configuration, scale-up, projected maintenance, and cost estimation are generally less established. Other reasons concern fundamental technical and economic issues. It is always more desirable to produce useful products at both electrodes, but, as a practical matter, it is often difficult to find operating conditions that are favorable for just one electrode reaction. More frequently, the products from the counterelectrode are discarded. For example, in the adiponitrile process the oxygen produced at the anode is vented. From an economic point of view, catalytic electrodes often require a larger capital investment per unit production rate. The requirement of a separator can also add significantly to the capital and operating costs; designing the process to avoid a separator, if possible, is extremely desirable. In a development program at Monsanto, redesign of the adiponitrile process in an undivided cell resulted in a reduction in energy consumption of over 50% and a reduction in capital costs of about 90%.

9.2.1   Electrochemical Reactors

Reactors for electrochemical processes fall into the same broad categories used in chemical engineering practice: batch, plug-flow, continuous-flow stirred tank, etc. The major difference with electrochemical processes is that potential provides an additional driving force, which must be taken into account in process design. The potential distribution in turn influences the current distribution. Although a uniform current distribution is desirable, it is difficult to attain uniformity because cell components such as corners and coplanar electrode-insulator arrangements are invariably present. Non-uniform current distribution is not in itself a serious problem, but varying overpotentials can lead to the production of undesirable side products. The loss in current efficiency and possible product contamination accompanying parasitic reactions can be serious problems.

Many electrolytic processes are carried out in the liquid phase. Compared with gas-phase reactions, mass transport limitations are often more severe. Several techniques are available to increase mass transport. Increasing fluid velocity is one means of enhancing mass transport. In addition to increasing the mean fluid velocity, promoting turbulence near the electrode surface also enhances mass transport. Protrusions or screens placed near the electrodes often serve this purpose.

Electrodes can be arranged to carry current in several different configurations (Fig. 9.7). In a monopolar configuration each electrode is connected separately to the power source and maintains the same

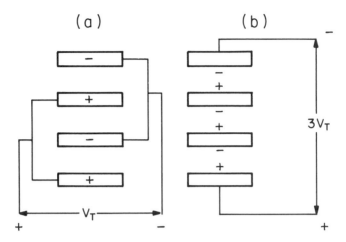

Figure 9.7: (a) Monopolar and (b) bipolar cell arrangements.

sign on both surfaces. In a bipolar arrangement the power source is connected to the outer two electrodes, and the two electrode surfaces on each interior electrode are of a different sign. The bipolar arrangement has simpler wiring, but there is greater chance for current leakage (shunt current) between cells. Although the total power input is about the same in both cases, the applied potential is higher in the bipolar case. It must equal the sum of the potential drops in all of the cells in the stack; however, the same current flows through each cell.

A component of electrochemical reactors rarely found in chemical reactors is the separator. Its function is to separate the constituents in the anode and cathode compartments. It serves to minimize homogeneous or back-reaction at the counterelectrode. Because separators add to capital cost and to operating cost through the additional ohmic drop incurred, they are undesirable. Some separators act as physical barriers and slow the mixing process; asbestos, porous glass, and plastics are in this class. Other materials are ion-selective. They permit the passage of ions of one sign only along with solvent. Mostly these materials are polymeric such as those based on perfluoropolymers with acidic side groups.

9.2.2   Chlor-alkali Processes

The commercial electrochemical synthesis of chlorine and sodium hydroxide from brine was developed about a century ago. Although the fundamental concept is unchanged, engineering improvements have made the process much more efficient. Because these chemicals are commodities, produced on the scale of 10^7 ton/y in the US, there is a large incentive to make small changes to reduce product cost. The evolutionary changes in cell design illustrate the application of technological advances to this established process.

The overall reaction is

$$2\,NaCl \;+\; 2\,H_2O \;=\; 2\,NaOH \;+\; Cl_2 \;+\; H_2 \qquad (9.16)$$

At the anode, chloride ions are oxidized.

$$2\,Cl^- \;\rightarrow\; Cl_2 \;+\; 2e \qquad (9.17)$$

At the cathode, hydroxide ions and hydrogen are produced.

$$2\,H_2O \;+\; 2e \;=\; H_2 \;+\; 2\,OH^- \qquad (9.18)$$

The standard reversible potential for the overall reaction is 2.19 V. Under typical operating conditions the reversible potential is about 2.23 V. The minimum energy requirement is equal to the change in free energy of the reaction, or

$$\begin{aligned}
\Delta G \;&=\; -nFE & (9.19)\\
&=\; -(2)(96500)(2.23) & (9.20)\\
&=\; -4.3 \times 10^5 \; J & (9.21)
\end{aligned}$$

This requirement translates to 1680 kW-h/ton of chlorine. For each ton of chlorine, 1.1 tons of sodium hydroxide and 28 kg of hydrogen are produced at 100% current efficiency.

Several routes have evolved for the production of chlorine. One of the oldest processes involves a mercury electrode. The reactions are

$$2\,Cl^- \;\rightarrow\; Cl_2 \;+\; 2e \qquad (9.22)$$
$$2\,Na^+ \;+\; 2\,Hg \;+\; 2e \;\rightarrow\; 2\,NaHg \qquad (9.23)$$

In a subsequent step the sodium amalgam is hydrolyzed as such:

$$2\,NaHg \; + \; 2\,H_2O \; \rightarrow \; H_2 \; + \; 2\,OH^- \; + \; 2\,Hg \qquad (9.24)$$

Cells based on this technology were formerly in widespread use, but in the 1950s several deaths were attributed to mercury poisoning from the facility at Minimata, Japan. As a result, mercury cells are being replaced in the US and Japan.

Early cells were designed with a physical barrier between the anolyte and catholyte sections; these cells with asbestos separators are known as diaphragm cells. Because the barrier is only a physical one, ions are free to pass between the compartments, and contamination of the sodium hydroxide with chloride occurs. The concentration of hydroxide that can be maintained in the catholyte is limited to about 10%. Higher concentrations allow significant diffusion to the anolyte where hydrolysis of the chlorine and additional oxygen production are favored; both of these effects reduce current efficiency. Early versions of the diaphragm cell used graphite anodes. Several engineering changes have significantly improved on this technology.

In the 1960s new anodes constructed from titanium, coated with ruthenium dioxide and other metal oxides, were developed. These anodes reduced chlorine overpotential while maintaining a high oxygen overpotential. Because these anodes were not consumed at an appreciable rate, they became known as dimensionally stable anodes. Such anodes have almost completely replaced carbon anodes in production cells.

The second major improvement was the development of a cation selective membrane. These membranes allow the passage of sodium ions, but prevent much of the diffusion of the chloride and hydroxide ions. Current membranes are made from perfluorinated polymers with carboxylic or sulfonic acid side groups. Production of 20-40% sodium hydroxide solutions is feasible. Because commercial grade sodium hydroxide is sold in concentrations of 50% or more, additional evaporation is still required, but the energy consumption for the evaporation is reduced.

These two developments have resulted in significant energy savings. It is possible to operate a membrane cell at twice the current density of a diaphragm cell, while maintaining about the same overall

Component	Membrane cell 450 mA/cm² (V)	Diaphragm cell 200 mA/cm² (V)
Thermodynamic potential	2.3	2.3
Anodic overpotential	0.05	0.03
Cathodic overpotential	0.3	0.2
Ohmic solution	0.3	0.5
Ohmic separator	0.4	0.5
Ohmic structural	0.3	0.2
Terminal voltage	3.65	3.73

Table **9.1**: Comparison of the components of cell voltage in diaphragm and membrane cells.

energy efficiency. An approximate breakdown of the components of cell potential appear in Table 9.1. Improved cathodes are being developed to reduce overpotential at that electrode. Raney nickel has been shown to reduce overpotential by 100 to 200 mV under current operating conditions.

Another possibility for reducing cell potential is to alter the thermodynamic component. Changes in operating temperature or pressure could only change the thermodynamic requirement by a few millivolts; however, a change in the overall reaction could significantly reduce the minimum energy required to carry out the decomposition. One such scheme is to reduce oxygen at the cathode.

$$\frac{1}{2} O_2 + H_2O + 2e = 2 OH^- \qquad (9.25)$$

The overall reaction is then

$$2 NaCl + \frac{1}{2} O_2 + H_2O = 2 NaOH + Cl_2 \qquad (9.26)$$

Under operating conditions the reversible potential is 1.1 V instead of the 2.3 V for conventional brine electrolysis. This scheme has been under development, but several practical difficulties have yet to be overcome. A significant problem is the sluggishness of the oxygen electrode. This same problem is faced in fuel cell technology. With an air electrode no hydrogen is produced; however, this drawback is not serious because the amount and purity of the hydrogen produced makes its use uneconomical in most cases.

9.2.3 Thermal Management

In systems where current density is relatively low, the amount of heat generated does not present a problem in terms of design. In high current density processes large quantities of heat can be produced primarily through irreversibilities in the cell. Although heat is also generated or consumed through reversible processes, this source tends to be smaller than irreversible heating.

Because the sources or sinks of heat are not uniformly distributed in a cell, a detailed temperature distribution cannot usually be obtained from a simple calculation. Another difficulty is that a cell is heterogeneous, and the components (electrodes, electrolyte, casing, separator) may all have significantly different thermal conductivities. A realistic simulation that could be used to determine a detailed temperature distribution would require the use of a numerical technique.

By contrast, an overall heat balance is relatively straightforward. Although the reversible heat can represent either heat absorbed or produced by a device, irreversible heating always represents a source of heat. Irreversibilities arise from overpotentials and ohmic drops. Overpotentials originate near electrode surfaces and cause local heating at those locations. Ohmic drop can occur in the electrolyte, separator, or in the structural components (cables, buses, and electrodes). The total heat production in an electrolyzer is the sum of the irreversible and reversible sources, or

$$-\dot{q} = I(V - E) - \frac{I}{nF} T\Delta S \qquad (9.27)$$

where V is the applied potential and E is the reversible cell potential. In a galvanic cell, terminal voltage is lower than the reversible potential; therefore, the sign of the first term must be reversed. Because the sign of the thermodynamic functions change sign with the direction of the reaction, the second term remains the same.

A positive value for $T\Delta S$ implies that an electrolyzer operating reversibly would need to absorb heat from the surroundings to maintain a constant temperature. This heat could be supplied by increasing the applied potential to a higher value so that the additional voltage would be converted to an equivalent amount of heat through irreversible losses. This higher potential is known as the thermoneutral potential E_{tn}. Use of the thermoneutral potential allows the combination of the two terms in Eq. 9.27.

$$- \dot{q} \; = \; I(V \; - \; E_{tn}) \qquad\qquad (9.28)$$

A calculation of the average operating temperature can be made by estimating a heat transfer coefficient from the device to the surroundings and making a heat balance.

9.3   Future Developments

The largest near-term potential applications of electrochemical technology are for utility load-leveling and for electric vehicle applications. Performance and cost factors still favor competing technologies. A more efficient electrolysis of water would make hydrogen a candidate for a primary gaseous energy currency, filling the role now occupied by natural gas. A significant advantage of hydrogen combustion is that it burns without the production of CO_2, which is linked to global warming. The use of utility load-leveling will be of greater importance if intermittent energy sources, such as wind and solar become more prevalent. Because nuclear power plants run more efficiently in the base-load mode, load-leveling will increase overall efficiency if a large number of nuclear units are installed. Several novel battery and fuel cell systems have been introduced in recent years, some aimed at large energy storage and conversion applications.

Fuel cells capable of operating at $1000°C$ have been under development. Laboratory cells have been fabricated from nickel anodes and strontium-doped lanthanum manganite cathodes. The electrolyte is a $20-\mu$ film of yttria-stabilized zirconia. At these temperatures both kinetic and ohmic losses are low, and high efficiencies are attainable. Current densities of 400 mA/cm^2 have been realized. These cells use carbon monoxide and water as fuel. Hydrogen is produced at the anode through the following reaction:

$$CO + H_2O \rightarrow CO_2 + H_2 \tag{9.29}$$

The hydrogen produced in this reaction is oxidized at the anode. At these temperatures oxygen ions are stable, and the individual electrode reactions are

$$O^{2-} + H_2 \rightarrow H_2O + 2e \tag{9.30}$$

$$\frac{1}{2} O_2 + 2e \rightarrow O^{2-} \tag{9.31}$$

A benefit of high-temperature operation is that the high quality waste heat can be used for additional thermal generation. The main disadvantage is that the cell is fragile and is susceptible to thermal stresses.

Batteries employing high energy-density materials have been under development for several decades. Those based on light metals with large, negative standard electrode potentials are particularly attractive. Most research efforts have been concentrated on aluminum, lithium, and sodium. Devices based on these systems are capable of specific energies on the order of 1000 W-h/kg. Because of the extreme activity of these metals, they cannot be recharged in aqueous electrolyte. Several modes of recharge have been considered. The Al/air battery can be discharged in basic solution, and the aluminum hydroxide product can be reprocessed chemically in a separate stage. Recharge of the Na/S battery from the molten state has been demonstrated; over 1000 cycles have been achieved in laboratory cells. Commercial versions of primary lithium cells with organic electrolyte have been available for a number of years. The overall reaction in thionyl chloride is

$$4\,Li + 2\,SOCl_2 \rightarrow 4\,LiCl + SO_2 + S \tag{9.32}$$

More recently a rechargeable version of the $Li/SOCl_2$ has been introduced.

The types of electrodes and electrode configurations are undergoing substantial change. A general problem with electrochemical reactors is to increase the space-time yield. For example, if we compare the throughput in conventional metal processing reactors with electrolytic processes, we find that the space-time yield for the conventional process is generally higher by at least an order of magnitude. The space-time yield for a typical pyrolytic process such as an iron blast furnace is on the order of 3 kg/m^3-min, but in an aluminum smelter it is only about 0.05 kg/m^3-min. Similar problems are evident in other areas of synthesis and energy production. Many electrochemical systems depend on porous electrodes to increase the available surface area per unit volume. Fluidized beds, common to many conventional processes, are being considered for metal refining and other electrochemical processes.

Significant research has been conducted on catalysis. The most important commercial breakthrough in recent times was the development of the dimensionally-stable anode for chlorine production. A great deal of effort is currently being directed toward improving the kinetics of the oxygen reduction reaction. This electrode reaction limits the utility of batteries and fuel cells employing an oxygen cathode. Although most electrodes are made from metals or carbon, other materials have recently been developed. Semiconducting electrodes have been studied in photoelectrochemical cells for potential use in solar energy conversion. Electrodes based on conducting organic materials have recently been devised. One type of electrode, composed of doped polyacetylene, may be effective in reducing device weight.

Many electrochemically-based devices or processes are still in the early stages of investigation. In the biological area numerous projects to monitor and control activity are underway. These projects range from monitoring blood sugar on enzyme electrodes to microelectronic triggers for the activation of nerve cells. Solid state fabrication will rely increasingly on electrochemical and photoelectrochemical techniques. Metals such as gold can be directly photoelectroplated onto semiconductor devices. With reductions in microcircuit dimensions reliability related to corrosion will become a greater problem. Because the distance between metal leads (runners) is small, high fields tend to promote corrosion on surfaces of microelectronic devices. Techniques for modifying the surface phases to make them less conductive and less subject to corrosion are being explored. Novel electrolytes are being investigated. Solid electrolytes, low-temperature molten salts, polymers and supercritical fluids have been considered in novel applications.

The increasing use of electricity as an energy source has several important consequences. Prototype electric vehicles have already been developed by major automobile companies. Since pollutants are more readily controlled from a single stationary source, the increasing use of electric vehicles will reduce urban air pollution. As hydrocarbon sources become scarcer, the premium on efficiency will stimulate additional research on electrochemical energy conversion and storage.

9.4 Problems

1. For a TRES operating between 1200 K and 400 K with methane as a fuel (see Fig. 9.6), estimate the best performance that could be obtained if all reactions ran reversibly. Express your answer in terms of the net heat added per unit of electrical energy produced; this quantity is sometimes expressed in Btu/kW-h and is called the heat rate.
b) Make a reasonable estimate of the losses at 100 mA/cm^2 and recalculate the heat rate.
c) Assume that the total capital cost, expressed in terms of a unit electrode area is $.02/cm^2, and assume a current density of 100 mA/cm^2. Estimate a capital cost ($/kW). The cost for a modern fossil fuel plant is roughly $3000/kW.

2. The availability of a material used in a device may limit its use in a large-scale application. Relatively scarce, noble, catalyst materials such as platinum may not be in sufficient supply for use in a particular application. Estimate the electric generating capacity (kW) that could be installed based on the use of phosphoric acid fuel cells with a catalyst loading of 0.1 mg/cm^2. Assume that the total production of platinum for one year (30,000 kg) were devoted to this purpose. Note that the total, installed, electric-generating capacity in the US is on the order of 500 million kW.

3. For spaceflights it would be more convenient to carry liquid methanol rather than hydrogen as the fuel for power generation from

a fuel cell. The standard reversible potential for the oxidation of liquid methanol to carbon dioxide and water is 1.214 V.

a) From thermodynamic considerations calculate the mass of reactants fed to a methanol fuel cell to produce the same amount of energy as could be obtained from 1 kg of hydrogen fuel cell reactants.

b) At a specified current density, would the relative specific energy (W-h/kg) of the methanol fuel cell be more or less favorable than that calculated in Part a? Explain your answer in qualitative terms.

4. High purity hydrogen is obtained commercially by the electrolysis of water in alkaline electrolyte. Because the enthalpy of reaction is greater than the Gibbs free energy of the reaction, a cell operating reversibly absorbs heat from the surroundings. At potentials greater than the reversible potential, heat is produced due to irreversibilities. At a sufficiently high potential the amount of heat produced exactly balances the heat absorbed by the cell; this potential is the thermoneutral potential. Estimate the thermoneutral potential for a water electrolysis cell operating at room temperature and one atmosphere pressure.

Bibliography

[1] E. J. Cairns and J. McBreen, *Advanced Batteries: Candidates for Electric Vehicle Propulsion*, General Motors Research Publication GMR-1871 (1975).

[2] R. A. Rapp in *Proceedings of Thermally Regenerative Electro-chemical Systems Workshop*, SERI/CP-234-1577, H. L. Chum, ed., pp. 177-183 (1975).

[3] S. Swann and R. C. Alkire, *Bibliography of Electro-organic Synthesis* (The Electrochemical Society: Princeton, 1980).

Appendices

Primary Sources of Data

Appendix B:

A. J. Bard, R. Parsons, and J. Jordan, *Standard Potentials in Aqueous Solution* (New York: Marcel Dekker, Inc., 1985).

Appendices C and D:

R. A. Robinson and R. H. Stokes, *Electrolyte Solutions* (London: Butterworths, 1959).

Appendix E:

J. R. Selman and C. W. Tobias, *Advances in Chemical Engineering*, **10**, 212 (1978).

A. Conversion Factors and Fundamental Constants

Conversion Factors

1 atm	$= 101.3$ kPa
1 atm	$= 1.013$ bar
1 atm	$= 14.70$ psi
1 A	$= 1$ C/s
1 A/cm^2	$= 929.0$ A/ft^2
1 Å	$= 10^{-8}$ cm
1 Btu	$= 1055$ J
1 cal	$= 4.184$ J
1 erg	$= 10^{-7}$ J
1 eV	$= 1.602 \times 10^{-19}$ J
1 F	$= 26.80$ A-h (Faraday)
1 F	$= 96{,}485$ C (Faraday)
1 F	$= 1$ C/V (farad)
1 H	$= 1$ ohm-s
1 mho	$= 1$ ohm^{-1}
1 mho	$= 1$ S (siemens)
1 ohm	$= 1$ V/C
1 W	$= 1$ J/s
1 W	$= 1.341 \times 10^{-3}$ hp
1 W-h	$= 3600$ J
1 V	$= 1$ J/s

Fundamental Constants

e	$= 1.602 \times 10^{-19}$ C
g	$= 980.7$ cm/s^2
k	$= 1.381 \times 10^{-23}$ J/K
R	$= 8.314$ J/mol-K
R	$= 1.987$ cal/mol-K
ϵ_0	$= 8.854 \times 10^{-12}$ F/m
μ_0	$= 4\pi \times 10^{-7}$ H/m

B. Standard Electrode Potentials

Reaction	E^0 (V)
$Ag^+ + e = Ag$	0.799
$AgCl + e = Ag + Cl^-$	0.222
$AgI + e = Ag + I^-$	-0.152
$Ag_2O + H_2O + 2e = 2\ Ag + 2\ OH^-$	0.342
$Al^{3+} + 3e = Al$	-1.66
$Au^+ + e = Au$	1.83
$Au^{3+} + 2e = Au^+$	1.401
$Au^{3+} + 3e = Au$	1.52
$Br_2(aq) + 2e = 2\ Br^-$	1.087
$Br_2(liq) + 2e = 2\ Br^-$	1.065
$Cd^{2+} + 2e = Cd$	-0.403
$Cl_2 + 2e = 2\ Cl^-$	1.358
$Co^{2+} + 2e = Co$	-0.277
$CO_2 + 2\ H^+ + 2e = HCOOH$	-0.199
$Cr^{2+} + 2e = Cr$	-0.557
$Cr^{3+} + 3e = Cr$	-0.744
$Cu^+ + e = Cu$	0.520
$Cu^{2+} + 2e = Cu$	0.340
$Cu(OH)_2 + 2e = Cu + 2\ OH^-$	-0.22
$2\ Cu(OH)_2 + 2e = Cu_2O + 2\ OH^-$	-0.08
$2\ D^+ + 2e = D_2$	-0.044
$F_2 + 2e = 2\ F^-$	2.87
$Fe^{2+} + 2e = Fe$	-0.44
$Fe^{3+} + e = Fe^{2+}$	0.771
$Fe(CN)_6^{3-} + e = Fe(CN)_6^{4-}$	0.361
$FeCl_2^+ + e = FeCl_2$	0.651
$Fe(OH)_3 + e = Fe(OH)_2 + OH^-$	-0.56
$2\ H^+ + 2e = H_2$	0.000
$2\ H_2O + 4e = 2\ OH^- + H_2$	-0.828

Reaction	E^0 (V)
$Hg^{2+} + 2e = Hg$	0.851
$Hg_2Cl_2 + 2e = 2\ Hg + 2\ Cl^-$	0.268
$HgO + H_2O + 2e = Hg + 2\ OH^-$	0.098
$I_2 + 2e = 2\ I^-$	0.535
$I_3^- + 2e = 3I^-$	0.534
$K^+ + e = K$	-2.925
$Li^+ + e = Li$	-3.045
$Mg^{2+} + 2e = Mg$	-2.375
$Mn^{2+} + 2e = Mn$	-1.029
$MnO_2 + 4\ H^+ + 2e = Mn^{2=} + 2\ H_2O$	1.208
$Ni^{2+} + 2e = Ni$	-0.23
$NiO_2 + 2\ H_2O + 2e = Ni(OH)_2 + 2\ OH^-$	0.49
$O_2 + 2\ H^+ + 2e = H_2O_2$	0.682
$O_2 + 4\ H^+ + 4e = 2\ H_2O$	1.229
$O_2 + 2\ H_2O + 4e = 4\ OH^-$	0.401
$O_3 + 2\ H^+ + 2e = O_2 + H_2O$	2.07
$Pb^{2+} + 2e = Pb$	0.126
$PbO + H_2O + 2e = Pb + 2\ OH^-$	-0.576
$PbO_2 + 4\ H^+ + 2e = Pb^{2+} + 2\ H_2O$	1.46
$PbO_2 + H_2O + 2e = PbO + 2\ OH^-$	0.28
$PbO_2 + SO_4^{2-} + 4\ H^+ + 2e = PbSO_4 + 2\ H_2O$	1.685
$3\ PbO_2 + 4\ H^+ + 4e = Pb_3O_4 + 2\ H_2O$	1.127
$PbSO_4 + 2e = Pb + SO_4^{2-}$	-0.356
$S + 2e = S^{2-}$	-0.508
$SO_4^{2-} + 4\ H^+ + 2e = H_2SO_3 + 2e = H_2SO_3 + H_2O$	0.20
$Ti(OH)_2^{2+} + 2\ H^+ + e = Ti^{3+} + 2\ H_2O$	0.1
$Zn^{2+} + 2e = Zn$	0.763
$ZnO_2^{2-} + 2\ H_2O + 2e = Zn + 4\ OH^-$	-1.216

C. Equivalent Conductances

Limiting equivalent ion conductivities λ^0 (cm^2/ohm-equiv) at selected temperatures:

Ion	0°C	25°C	100°C
Ag^+	33.1	61.9	175
Ca^{2+}	31.2	59.5	180
H^+	225	349.8	630
K^+	40.7	73.5	195
Li^+	19.4	38.7	115
Na^+	26.5	50.1	145
Br^-	42.6	78.1	-
Cl^-	41.0	76.4	212
I^-	41.4	76.8	-
NO_3^-	40.0	71.5	195
OH^-	105	198.3	450
SO_4^{2-}	41	80.0	260

Equivalent conductivity Λ (cm^2 ohm^{-1} $equiv^{-1}$) of typical electrolytes at 25°C at various concentrations (mol/L):

c	NaCl	KCl	$BaCl_2$	$LaCl_2$
0	126.4	149.9	140.0	146.0
0.0005	124.5	147.8	134.3	135.2
0.001	123.7	147.0	132.3	131.2
0.005	120.6	143.6	124.0	118.1
0.01	118.5	141.3	119.1	111.2
0.05	111.1	133.4	105.2	95.0
0.1	106.7	129.0	98.7	87.9
0.5	93.6	117.3	80.6	66.7
1.0	85.8	111.9	69.0	51.2

D. Activity Coefficients of Electrolytes at 25°C

Mean molal activity coefficients at various concentrations:

m	$CuSO_4$	HCl	KCl	NaCl	NaOH
0.1	0.150	0.796	0.770	0.778	0.764
0.2	0.104	0.767	0.718	0.735	0.725
0.3	0.083	0.756	0.688	0.710	0.706
0.4	0.070	0.755	0.666	0.693	0.695
0.5	0.062	0.757	0.649	0.681	0.688
0.8	0.048	0.783	0.618	0.662	0.677
1.0	0.042	0.809	0.604	0.657	0.677
1.6		0.916	0.580	0.657	0.690
2.0		1.01	0.573	0.668	0.707
2.5		1.147	0.569	0.688	0.741
3.0		1.316	0.569	0.714	0.782
4.0		1.762	0.577	0.783	0.901
5.0		2.38		0.874	1.074
6.0		3.22		0.986	1.296

m	HCl	LiCl	NaOH	KOH	NH_4NO_3
7	4.37	3.71	1.60	2.80	0.261
8	5.90	5.10	2.00	3.66	0.245
9	7.94	6.96	2.54	4.72	0.232
10	10.44	9.40	3.22	6.05	0.221
12	17.25	16.41	5.18	10.2	0.202
14	27.3	26.2	8.02	15.4	0.187
16	42.4	37.9	11.55	23.9	0.174
18		49.9	15.37		0.163
20		62.4	19.28		0.153

E. Mass Transport Correlations

System/ Range	Parameters	Correlation
LAMINAR, FORCED CONVECTION		
Rotating disk $Re < 2 \times 10^5$	ω = rotation rate r = disk radius $Re = r^2\omega/\nu$ $Sh = i_l r/(nFDc)$	$Sh = 0.62\ Re^{1/2}Sc^{1/3}$
Rotating hemisphere $Re < 15{,}000$	r = radius of sphere Other parameters same as disk.	$Sh = 0.474\ Re^{1/2}Sc^{1/3}$
Rotating cone $Re < 25{,}000$ $25° < \theta < 60°$	r = radius of base θ = cone angle (half opening) Other parameters same as disk.	$Sh = 0.62\ (\sin\theta)^{1/2}\ Re^{1/2}Sc^{1/3}$

System/ Range	Parameters	Correlation
LAMINAR, FORCED CONVECTION		
Annular duct $1.7 \times 10^4 < ReSc$ $d_e/L < 9 \times 10^7$	$d_e = r_0 - r_i$ $r_i/r_0 = 0.5$ $r_0 = $ outer radius $Re = v_{avg} d_e/\nu$ $Sh = i_l d_e/(nFDc)$	$Sh = 1.94(ReSc\ d_e/L)^{1/3}$ $r_i = $ inner radius
Flow channel $75 < Re < 7000$ $600 < Sc < 12{,}000$ $0.05 < d_e/L < 20$	$h = $ electrode separation $Re = v_{avg} d_e/\nu$ $d_e = $ equivalent diameter $Sh = i_l d_e/(nFDc)$	$Sh = 1.85(ReSc\ d_e/L)^{1/3}$

System/ Range	Parameters	Correlation
LAMINAR, FREE CONVECTION		
Vertical plate $5 \times 10^6 < GrSc < 5 \times 10^{12}$ $600 < Sc < 12{,}000$	$L = $ height $Gr = g\Delta\rho L^3/\rho\nu^2$	$Sh = 0.67(GrSc)^{1/4}$
Horizontal cylinder $2.3 \times 10^6 < GrSc < 10^9$	$d = $ cylinder diameter $Sh = i_l d/(nFDc)$ $Gr = g\Delta\rho d^3/\rho\nu^2$	$Sh = 0.23(GrSc)^{0.3}$
Sphere $2.3 \times 10^8 < GrSc < 1.5 \times 10^{10}$	$d = $ sphere diameter	$Sh = 2 + 0.59(GrSc)^{0.25}$
TURBULENT, FREE CONVECTION		
Vertical cylinder $2.9 < GrSc < 400$ $2500 < Sc < 3800$ $3700 < L/r < 78{,}000$	$r = $ cylinder radius $Gr = g\Delta\rho r^3/\nu^2\rho$ $L = $ cylinder length	$Sh = 0.57(GrSc)^{0.11}$
Vertical plate $4 \times 10^{13} < GrSc < 10^{15}$	$L = $ height of plate	$Sh = 0.31(GrSc)^{0.28}$

System/ Range	Parameters	Correlation
	TURBULENT, FORCED CONVECTION	
Rotating disk $8.9 \times 10^5 < GrSc <$ 1.18×10^7	ω = rotation rate $Re = \omega r^2/\nu$	$Sh = 0.0117\ Re^{0.896}Sc^{0.249}$
Rotating cylinder $10^3 < Re < 10^5$ $835 < Sc < 11{,}490$ $0.093 < d_i/d_o < 0.83$	ω = rotation rate d_i = inner diameter d_o = outer diameter $Sh = i_l d_i/(nFDc)$ $Re = \omega d_i^2/\nu$	$Sh = 0.0791\ Re^{0.7}Sc^{0.356}$
Annular duct or Tube $(r_i = 0)$ $2100 < Re < 30{,}000$ $r_i/r_0 = 0.5$	$d_e = r_0 - r_i$ r_i = inner radius r_0 = outer radius $Re = v_{avg}d_e/\nu$ $Sh = i_l d_e/(nFDc)$	$Sh = 0.023\ Re^{0.8}Sc^{1/3}$
Flow channel $12{,}000 < Re < 125{,}000$ $1200 < Sc < 25{,}100$	d_e = equivalent diameter $Sh = i_l d_e/(nFDc)$	$Sh = 0.01\ Re^{0.92}Sc^{0.336}$

F. Computer Program for a One-dimensional Cell

The potential distribution in a one-dimensional cell is known from elementary theory to be linear. For a secondary or tertiary current distribution, the potential adjacent to the electrode (ϕ_0) is different from the electrode potential (V) and must be calculated by an analytical or numerical technique. If a simple, linear overpotential expression is assumed at one or both electrodes, an analytical expression for ϕ_0 can be derived as shown in Sec. 7.3.3. Because both analytical and numerical solutions are available for the secondary potential distribution and current, we can compare them and examine the stability of the finite-difference method.

The specific problem considered here is the one-dimensional electrolytic cell having overpotential that is linearly related to the current density at the cathode. The anode is assumed to be unpolarized. We can use the finite-difference method and develop a Fortran program to calculate ϕ_0, i, and η_{sc}.

Several of the variables are set within the program and can be changed only by altering the program. The linear polarization parameter, electrolyte conductivity, and node spacing can be varied by changing the parameters in an input file. The principal variables are listed in Table 1.

A brief explanation of the Fortran program ONED follows. It was written using the standard features of Fortran 77. The numbers within parentheses refer to the listing numbers in the left margin of each statement.

The three parameters (B, CON, and H) are entered (7), and all other parameters are set in a DATA statement (8). The number of nodes is calculated by dividing the cell width W by the node spacing (17); a small increment is added to W to prevent round-off errors. All interior nodes are initially set (20) to a potential corresponding to the average potential (1.5 V). For linear polarization, the Wagner number (24) is CON*B/W, where W = 1.

Iteration begins (27) and new values of the potentials are calculated (31) from Laplace's equation using the finite-difference approx-

Fortran variable	Description
B	Linear polarization parameter $(\text{V-cm}^2/\text{A})$ in $\eta = Bi$
CDA	Current density at the anode
CDC	Current density at the cathode
CON	Electrolyte conductivity $(\text{ohm}^{-1}\text{-cm}^{-1})$
EN	Relative difference in potential between iterations
ERR	Convergence criterion (10^{-5})
ETA	Surface overpotential (V)
H	Spacing between nodes (cm)
ITR	Number of iterations
ITMAX	Maximum number of iterations (1500)
NODE	Number of nodes
NUN	Number of unconverged nodes
PN	Potential at most recent iteration
P(I)	Potential at node i
RBPE	Relative boundary potential error
VA	Anode potential (V)
VC	Cathode potential (V)
W	Width of cell (1 cm)

Table 1: Fortran variables used in the program ONED.

imation derived in Eq. 7.42. The relative change in potential is the relative difference between the new value and the previous value at each iteration (32). If the relative change exceeds the specified error criterion ERR, then the counter NUN is incremented (33). After the error calculation, the old value is replaced with the new potential value (34).

Current densities are calculated at the anode and cathode through numerical differentiation (37-38) of the potential using Eqs. 7.34-7.35. Overpotential at the cathode is calculated from the linear polarization equation (40). The potential adjacent to the cathode is obtained by subtracting the overpotential from the cathode potential (41). The relative change in potential adjacent to the cathode is calculated (42) by the same method that was used for the interior potentials (32).

The progress of the iteration scheme is displayed on the default device (usually a terminal) every 20 iterations (44-46).

If all of the potential changes are within the error criteria, then convergence is assumed (47-50), and the results are printed to a file (51-60). If one of the criteria is not met, then iteration continues. If the number of iterations exceeds the specified maximum (1500 iterations), then a message indicating this condition is printed (200). The analytical solutions for the potential adjacent to the cathode and the current density are calculated and printed (61-66).

The sample output shows the results of a simulation using a conductivity of 1 ohm^{-1}-cm^{-1}, a polarization parameter of 0.1 A/(V/cm^2), and a node spacing of 0.1 cm. From the numerical solution, the potential adjacent to the cathode is 1.091 V, which implies that the overpotential is 91 mV. The current densities at the anode and cathode should be the same, but because of the small errors in potentials, the currents are different from the analytical values by a fraction of a percent. The analytical value of ϕ_0 is the same as the numerical value to 3 decimals. This degree of convergence required 80 iterations. The number of iterations required increases slowly until a Wagner number of 0.4 is reached. This algorithm becomes unstable at higher Wagner numbers.

```
0001   C   ONED.FOR VER 1.0
0002           DIMENSION P(0:101)
0003           OPEN(UNIT=1,FILE='ONED.DAT',STATUS='OLD')
0004           OPEN(UNIT=2,FILE='ONED.OUT',STATUS='NEW')
0005           WRITE (2,3)
0006     3     FORMAT('     POTENTIAL DISTRIBUTION FOR A ONE-DIMENSIONAL CELL'//)
0007           READ (1,*) B, CON, H
0008           DATA ERR, ITMAX, W, VA, VC / 1E-5, 1500, 1., 2., 1./
0009           WRITE (2,4) B, CON, H
0010     4     FORMAT('   LINEAR POLARIZATION PARAMETER= '1PE11.2' V/(A/CM^2)'/
0011          $'   ELECTROLYTE CONDUCTIVITY= '1PE11.2' (OHM-CM)^-1'/
0012          $'   NODE SPACING= '0PF8.3' CM')
0013           WRITE (2,5) VA,VC
0014     5     FORMAT('   ANODE POTENTIAL= 'F4.1' V'/
0015          $'   CATHODE POTENTIAL= 'F4.1' V'/)
0016   C   CALCULATE NUMBER OF NODES
0017           NODE=(W+1.E-6)/H
0018   C   INITIALIZE VARIABLES
0019           DO 10 I=1,NODE-1
0020           P(I)=(VA+VC)/2.
0021    10     CONTINUE
0022           P(0)=VC
0023           P(NODE)=VA
0024           WRITE(2,20) CON*B
0025    20     FORMAT('  WAGNER NUMBER= 'F6.3/)
0026   C   BEGIN ITERATION
0027   100     CONTINUE
0028           NUN=0
0029           ITR=ITR+1
0030           DO 110 I=1,NODE-1
0031           PN=(P(I-1)+P(I+1))/2.
0032           EN=ABS(PN-P(I))/(ABS(PN)+1.E-6)
0033           IF(EN.GT.ERR)NUN=NUN+1
0034           P(I)=PN
0035   110     CONTINUE
0036   C CALCULATE CURRENT DENSITIES
0037           CDC=-CON*(P(2)-P(1))/H
0038           CDA=-CON*(P(NODE-1)-P(NODE))/H
0039   C CALCULATE OVERPOTENTIAL AT CATHODE
0040           ETA=B*CDC
0041           PN=VC-ETA
0042           RBPE=(ABS(PN-P(0)))/(ABS(PN)+1.E-6)
0043           P(0)=PN
0044           IF(ITR.EQ.1) WRITE(*,115)ITR,P(0),CDC
0045           IF(MOD(ITR,20).EQ.0) WRITE(*,115)ITR,P(0),CDC
0046   115     FORMAT('  ITERATION 'I4'  P0 'F8.4'  CURRENT DENSITY ' F8.4)
0047   C   CHECK CONVERGENCE
0048           IF(ITR.GT.ITMAX) GO TO 200
0049           IF(NUN.NE.0) GO TO 100
0050           IF(RBPE.GT.10*ERR) GO TO 100
```

```
0051          WRITE(2,117)
0052     117  FORMAT(15X,'NUMERICAL SOLUTION')
0053          WRITE(2,120)
0054     120  FORMAT(20X,'POTENTIALS'/)
0055          WRITE(2,130) (P(I), I=0,NODE)
0056     130  FORMAT(2X,11(F6.3,1X))
0057          WRITE(2,140) CDC,CDA
0058     140  FORMAT(/2X,'CURRENT DENSITY: 'F8.4' CATHODE'4X,F8.4' ANODE')
0059          WRITE(2,145) ITR
0060     145  FORMAT('  ITERATIONS = 'I5)
0061          WRITE(2,150)
0062     150  FORMAT(/20X,'ANALYTICAL SOLUTION'/)
0063          PC=(VC+B*CON*VA/W)/(1+B*CON/W)
0064          CD=-CON*(VA-PC)/W
0065          WRITE(2,160)PC,CD
0066     160  FORMAT('  CATHODE POTENTIAL 'F6.3'  CURRENT DENSITY 'F8.4/)
0067          GO TO 220
0068     200  CONTINUE
0069          WRITE(2,210)
0070     210  FORMAT('  NO CONVERGENCE')
0071     220  CONTINUE
0072          END
```

POTENTIAL DISTRIBUTION FOR A ONE-DIMENSIONAL CELL

LINEAR POLARIZATION PARAMETER=   1.00E-01 V/(A/CM^2)
ELECTROLYTE CONDUCTIVITY=   1.00E+00 (OHM-CM)^-1
NODE SPACING=   0.100 CM
ANODE POTENTIAL=   2.0 V
CATHODE POTENTIAL= 1.0 V

WAGNER NUMBER=   0.100

                    NUMERICAL SOLUTION
                       POTENTIALS

  1.091   1.182   1.273   1.363   1.454   1.545   1.636   1.727   1.818   1.909   2.000

CURRENT DENSITY: -0.9087 CATHODE   0.9095 ANODE
ITERATIONS =   80

                    ANALYTICAL SOLUTION

CATHODE POTENTIAL   1.091   CURRENT DENSITY   -0.9091

G. Computer Program for a Two-dimensional L-cell

The one-dimensional cell simulation described in Appendix F is an example of the application of numerical methods to the calculation of potential distribution. In that case the accuracy of the numerical method could be checked through comparison with the analytical solution. For most two-dimensional cells, an analytical solution is not available, and we are forced to rely on numerical simulation.

The following computer program was developed for the L-shaped cell described in Sec. 7.3.3. The current and potential distributions are calculated using the finite-difference approach. Unlike the one-dimensional cell described in Appendix F, there is no analytical solution to this problem.

The L-shaped cell in the simulation is depicted in Fig. 7.17, where the cathode is the rectangular portion in the lower left-hand corner. This program is constructed so that geometric ratios, Tafel parameters, and electrolyte conductivity can be varied. Principal variables are listed in Table 1.

An explanation of the program follows. The numbers within parentheses refer to the listing numbers in the left margin of each statement. The geometric, polarization, and conductivity data are read in (6-20). The polarization equation is the Tafel expression in the form $\eta_s = B \log(i/i_0)$. The numbers of nodes are computed by dividing the cell dimensions by the node spacing (23-26). For an initial estimate all nodes are set to the average of the applied electrode potentials (31-36). Electrode potentials are set equal to the specified potentials (38-46).

Iteration begins (48) and the interior potentials are calculated (52-60) from the two-dimensional analog of Eq. 7.42.

$$\phi_0 = \frac{\phi_{i-1,j} + \phi_{i}, j + 1 + \phi_{i+1,j} + \phi_{i,j-1}}{4}$$

Convergence can be hastened by using an overrelaxation scheme. Rather than simply replacing the previous potential with the updated

Fortran variable	Description
AVP	Average applied potential
B	Tafel polarization parameter (V)
CAH	Cathode height (cm)
CAW	Cathode width (cm)
CD(I,J)	Current density (A/cm^2)
CD0	Exchange current density (A/cm^2)
CH	Cell height (cm)
CON	Electrolyte conductivity (ohm^{-1}-cm^{-1})
CW	Cell width (cm)
EN	Relative difference in potential between iterations
ERR	Convergence criterion (10^{-5})
ETA	Surface overpotential (V)
H	Spacing between nodes (cm)
I	Column number
ICAW	Number of columns for the cathode
ICW	Number of columns for the cell
ITR	Number of iterations
ITMAX	Maximum number of iterations (1500)
J	Row number
JCAH	Number of rows for the cathode
JCH	Number of rows for the cell
NUN	Number of unconverged nodes
P(I,J)	Potential (V)
PN	Potential at most recent iteration
RBPE	Relative boundary potential error
VA	Anode potential (V)
VC	Cathode potential (V)
WA	Wagner number

Table 1: Principal variables used in the program COR.

value, the new value can be weighted more strongly.

$$\phi^{r+1} = \phi^r + F(\phi^{r+1} - \phi^r)$$
$$F\phi^{r+1} + (1 - F)\phi^r$$

where the superscript r represents the previous iteration and $r + 1$ represents the updated value. A value of $F = 1.85$ has been shown empirically to be a reasonable value for many problems of this type and is used in this program (63).

The image point method is used in the computation of potentials at insulated surfaces. Because there is no normal potential gradient at an insulated surface, the potential within the insulated surface (image point) is assumed to be at the same potential as one that is in a corresponding position in the electrolyte. For example, if the node at *i-1* is within an insulator, then the finite-difference equation is

$$\phi_0 = \frac{\phi i, j + 1 + 2\phi_{i+1,j} + \phi_{i,j-1}}{4}$$

All potentials at insulators are calculated using analogous formulations (67-79).

Information on intermediate results is displayed on the screen every 50 iterations (80-81). If all of the potentials on the interior nodes are within the convergence criterion then the current densities are recomputed (84). Alternatively, every five iterations, the current densities are recomputed by a two-point numerical differentiation formula (89-99).

The overpotential at each node on the cathode is computed from the Tafel expression (104). The relative changes in the cathode potentials are summed (106, 113). Values of the cathode potentials are underrelaxed (107, 114) such that $F < 0.05$. This simple algorithm increases the stability and range of Wagner numbers for which convergence can be achieved. If the convergence criteria are met (117-119), the results are printed (121-146); otherwise, iteration continues until the maximum specified number of iterations is exceeded.

The sample output is shown for a cell that is 1-cm square with a cathode that is 0.5-cm square. If the number of columns is less than 12, the potential map gives a coherent picture of the distribution. To distinguish the cathode nodes from interior nodes, we have arbitrarily designated the cathode with -9.9 symbols. The potentials on the cathode are described by the values in row 5 and in column 5 adjacent to the -9.9 symbols. Because no polarization is specified for the anode, its potential is held at a constant value of 2 V. The potential on the cathode is more positive in areas of highest current density where overpotential is greatest.

The current densities on the anode (row 10) are indicated from column 0 to column 11. They are highest in the area where the anode is immediately opposite the cathode. Current densities on the cathode are shown in row 5 from column 1 to column 6 and in column 5 from row 0 to row 4. The current density on the cathode is highest near the corner, but small variations in that area may be due to inaccuracies in numerical differentiation.

Calculation of the Wagner number is based on the anode-cathode separation and on the current density in column 0, away from the corner. In this example the Wagner number is 1.4. Convergence to the specified level was attained after 610 iterations in the potential loop.

Both of these programs are available on disk from the author for $10; they can be copied for class use or other purposes. A more general program to treat arbitrary two-dimensional cells with a range of polarization expressions and multiple electrodes is also available from the author.

```
0001    C COR.FOR VER 1.0
0002          DIMENSION P(0:50,0:50),CD(0:50,0:50)
0003          CHARACTER*80 TITLE
0004          OPEN(UNIT=1,FILE='COR.DAT',STATUS='OLD')
0005          OPEN(UNIT=2,FILE='COR.OUT',STATUS='NEW')
0006          READ(1,10) TITLE
0007          WRITE(2,10) TITLE
0008    10    FORMAT(A)
0009          READ(1,*) VA,VC,CON
0010          WRITE(2,20) VA,VC,CON
0011    20    FORMAT(/' ANODE POTENTIAL= 'F7.3' CATHODE POTENTIAL= ' F7.3/
0012          $' CONDUCTIVITY (OHM-1 CM-1) '1PE11.3/)
0013          READ(1,*) B,CD0
0014          WRITE(2,30) B,CD0
0015    30    FORMAT(' TAFEL SLOPE= 'F7.3' EXCHANGE CURRENT= '1PE11.3/)
0016          READ(1,*) CAW,CW,CAH,CH,H
0017          WRITE(2,40) CAW,CW,CAH,CH,H
0018    40    FORMAT(' CATHODE WIDTH= 'F7.3' CELL WIDTH= 'F7.3/
0019          $' CATHODE HEIGHT= 'F7.3' CELL HEIGHT= 'F7.3/
0020          $' MESH SPACING= 'F7.3/)
0021          DATA ERR, ITMAX/1.E-5, 1500/
0022    C CALCULATE NUMBER OF NODES CORRESPONDING TO CELL DIMENSIONS
0023          ICAW=(CAW+1.E-6)/H
0024          ICW=(CW+1.E-6)/H
0025          JCAH=(CAH+1.E-6)/H
0026          JCH=(CH+1.E-6)/H
0027    C INITIALIZE VARIABLES
0028          ITR=0
0029          RBPE=1.
0030    C SET ALL INTERIOR POTENTIALS TO THE AVERAGE POTENTIAL
0031          AVP=(VA+VC)/2.
0032          DO 60 I=0,ICW
0033          DO 50 J=0,JCH
0034          P(I,J)=AVP
0035    50    CONTINUE
0036    60    CONTINUE
0037    C SET ELECTRODES TO SPECIFIED POTENTIAL
0038          DO 70 J=0,JCAH
0039          P(ICAW,J)=VC
0040    70    CONTINUE
0041          DO 80 I=0,ICAW
0042          P(I,JCAH)=VC
0043    80    CONTINUE
0044          DO 90 I=0,ICW
0045          P(I,JCH)=VA
0046    90    CONTINUE
0047    C BEGIN ITERATION
0048    100   CONTINUE
0049          ITR=ITR+1
0050          NUN=0
```

```
0051    C CALCULATE ALL INTERIOR POTENTIALS
0052          DO 120 I=1,ICW-1
0053          DO 110 J=JCAH+1,JCH-1
0054          PN=(P(I-1,J)+P(I,J+1)+P(I+1,J)+P(I,J-1))/4.
0055    C CHECK ERROR BETWEEN ITERATIONS
0056          EN=ABS(PN-P(I,J))/(ABS(PN)+1.E-10)
0057          IF(EN.GT.ERR)NUN=NUN+1
0058          P(I,J)=1.85*PN-.85*P(I,J)
0059      110 CONTINUE
0060      120 CONTINUE
0061          DO 140 I=ICAW+1,ICW-1
0062          DO 130 J=1,JCAH
0063          P(I,J)=1.85*(P(I-1,J)+P(I,J+1)+P(I+1,J)+P(I,J-1))/4.-.85*P(I,J)
0064      130 CONTINUE
0065      140 CONTINUE
0066    C CALCULATE POTENTIALS ON INSULATED SURFACES
0067          I=0
0068          DO 150 J=JCAH+1,JCH-1
0069          P(I,J)=1.85*(P(I,J+1)+2.*P(I+1,J)+P(I,J-1))/4.-.85*P(I,J)
0070      150 CONTINUE
0071          I=ICW
0072          DO 160 J=1,JCH-1
0073          P(I,J)=1.85*(2.*P(I-1,J)+P(I,J+1)+P(I,J-1))/4.-.85*P(I,J)
0074      160 CONTINUE
0075          J=0
0076          DO 170 I=ICAW+1,ICW-1
0077          P(I,J)=1.85*(P(I+1,J)+P(I-1,J)+2.*P(I,J+1))/4.-.85*P(I,J)
0078      170 CONTINUE
0079          P(ICW,0)=1.85*(2*P(ICW-1,0)+2*P(ICW,1))/4.-.85*P(I,J)
0080          IF(MOD(ITR,50).EQ.0)WRITE(*,180)ITR,NUN,P(ICAW,JCAH),CD(ICAW,JCAH)
0081      180 FORMAT(' ITR= 'I4'  UNCONV = 'I3'  P(COR)= 'F6.2'  CD(COR)=' F6.2)
0082    C CALCULATE CURRENT DENSITIES
0083          IF(ITR.GT.ITMAX) GO TO 200
0084          IF(NUN.EQ.0.AND.B.LT.1.E-6) GO TO 200
0085          IF(MOD(ITR,5).EQ.0.AND.B.GT.1.E-6) GO TO 200
0086          GO TO 100
0087      200 CONTINUE
0088    C CALCULATE CURRENT DENSITIES
0089          DO 210 I=0,ICW
0090          CD(I,JCH)=-CON*(P(I,JCH-1)-P(I,JCH))/H
0091      210 CONTINUE
0092          DO 220 I=0,ICAW-1
0093          CD(I,JCAH)=-CON*(P(I,JCAH+1)-P(I,JCAH))/H
0094      220 CONTINUE
0095          DO 230 J=0,JCAH-1
0096          CD(ICAW,J)=-CON*(P(ICAW+1,J)-P(ICAW,J))/H
0097      230 CONTINUE
0098          CD(ICAW,JCAH)=-CON*(P(ICAW+1,JCAH+1)-P(ICAW,JCAH))/(1.414*H)
0099          IF(NUN.EQ.0.AND.B.LT.1.E-6) GO TO 300
```

```
0100    C CALCULATE OVERPOTENTIALS AT CATHODE
0101          TRBE=0
0102          J=JCAH
0103          DO 240 I=0,ICAW
0104          ETA=B*LOG10(ABS(CD(I,J))/CD0)
0105          PN=VC+ETA
0106          TRBE=(ABS(PN-P(I,J)))/(ABS(PN)+1.E-10)+TRBE
0107          P(I,J)=P(I,J)+.05*(PN-P(I,J))*(ITMAX-ITR+1)/REAL(ITMAX)
0108    240   CONTINUE
0109          I=ICAW
0110          DO 250 J=0,JCAH-1
0111          ETA=B*LOG10(ABS(CD(I,J))/CD0)
0112          PN=VC+ETA
0113          TRBE=(ABS(PN-P(I,J)))/(ABS(PN)+1.E-10)+TRBE
0114          P(I,J)=P(I,J)+.05*(PN-P(I,J))*(ITMAX-ITR+1)/REAL(ITMAX)
0115    250   CONTINUE
0116          RBPE=TRBE/(ICAW+JCAH-1)
0117          IF(ITR.GT.ITMAX) GO TO 300
0118          IF(NUN.NE.0) GO TO 100
0119          IF(RBPE.GT.50.*ERR) GO TO 100
0120    C PRINT RESULTS
0121    300   CONTINUE
0122          WRITE(2,310)
0123    310   FORMAT(/'                    POTENTIALS'//' ROW'/)
0124          DO 330 J=JCH,0,-1
0125          IF(J.LT.JCAH) THEN
0126          DO 315 I=0,ICAW-1
0127          P(I,J)=-9.9
0128    315   CONTINUE
0129          ENDIF
0130          WRITE(2,320) J,(P(I,J),I=0,ICW)
0131    320   FORMAT(I4,2X,(11(F4.1,1X)))
0132    330   CONTINUE
0133          WRITE(2,340)
0134    340   FORMAT(/'                 CURRENT DENSITIES'//)
0135          WRITE(2,350) JCH,(CD(I,JCH),I=0,ICW)
0136    350   FORMAT(/' ROW= 'I4/ (5(1PE10.2,1X)))
0137          WRITE(2,350) JCAH,(CD(I,JCAH),I=0,ICAW)
0138          WRITE(2,360) ICAW,(CD(ICAW,J),J=0,JCAH-1)
0139    360   FORMAT(/' COLUMN= 'I4/ (5(1PE10.2,1X)))
0140          WA=CON*B/((CH-CAH)*CD(0,JCH))
0141          WRITE(2,365) WA
0142    365   FORMAT(/ ' WAGNER NUMBER= 'F5.1)
0143          IF(ITR.LE.ITMAX) THEN
0144          WRITE(2,370)ITR
0145    370   FORMAT(' SOLUTION CONVERGED AFTER 'I4' ITERATIONS')
0146          ELSE
0147          WRITE(2,380)ITR
0148    380   FORMAT(' NO CONVERGENCE AFTER 'I4' ITERATIONS')
0149          ENDIF
0150          END
```

TEST CELL 1 CM SQUARE

ANODE POTENTIAL=   2.000 CATHODE POTENTIAL=   1.000
CONDUCTIVITY (OHM-1 CM-1)   1.000E-01

TAFEL SLOPE=   0.400 EXCHANGE CURRENT=   1.000E-03

CATHODE WIDTH=   0.500 CELL WIDTH=   1.000
CATHODE HEIGHT=   0.500 CELL HEIGHT=   1.000
MESH SPACING=   0.100

                 POTENTIALS

ROW

10   2.0   2.0   2.0   2.0   2.0   2.0   2.0   2.0   2.0   2.0   2.0
 9   1.9   1.9   1.9   1.9   1.9   1.9   1.9   1.9   2.0   2.0   2.0
 8   1.9   1.9   1.9   1.9   1.9   1.9   1.9   1.9   1.9   1.9   1.9
 7   1.8   1.8   1.8   1.8   1.8   1.8   1.8   1.9   1.9   1.9   1.9
 6   1.8   1.8   1.8   1.8   1.8   1.8   1.8   1.8   1.8   1.8   1.8
 5   1.7   1.7   1.7   1.7   1.7   1.7   1.7   1.8   1.8   1.8   1.8
 4 -9.9 -9.9 -9.9 -9.9 -9.9 -9.9   1.7   1.7   1.7   1.7   1.8
 3 -9.9 -9.9 -9.9 -9.9 -9.9 -9.9   1.7   1.7   1.7   1.7   1.7
 2 -9.9 -9.9 -9.9 -9.9 -9.9 -9.9   1.6   1.7   1.7   1.7   1.7
 1 -9.9 -9.9 -9.9 -9.9 -9.9 -9.9   1.6   1.7   1.7   1.7   1.7
 0 -9.9 -9.9 -9.9 -9.9 -9.9 -9.9   1.6   1.7   1.7   1.7   1.7

                 CURRENT DENSITIES

ROW

10   5.77E-02   5.76E-02   5.72E-02   5.66E-02   5.56E-02
5.42E-02   5.22E-02   5.02E-02   4.84E-02   4.72E-02
4.68E-02
 5   -5.92E-02   -5.92E-02   -5.94E-02   -5.98E-02   -6.04E-02
-5.87E-02

COLUMN

 5   -3.22E-02   -3.28E-02   -3.44E-02   -3.75E-02   -4.27E-02

 WAGNER NUMBER=   1.4
SOLUTION CONVERGED AFTER   610 ITERATIONS

Index